Spark-Plasma Sintering and Related Field-Assisted Powder Consolidation Technologies

Special Issue Editor
Eugene A. Olevsky

MDPI

Special Issue Editor
Eugene A. Olevsky
College of Engineering San Diego State University San Diego
USA

Editorial Office
MDPI AG
St. Alban-Anlage 66
Basel, Switzerland

This edition is a reprint of the Special Issue published online in the open access journal *Materials* (ISSN 1996-1944) from 2015–2016 (available at: http://www.mdpi.com/journal/materials/special_issues/Spark_Plasma_Sintering).

For citation purposes, cite each article independently as indicated on the article page online and as indicated below:

Author 1; Author 2; Author 3 etc. Article title. *Journal Name*. **Year**. Article number/page range.

ISBN 978-3-03842-382-9 (Pbk)
ISBN 978-3-03842-383-6 (PDF)

Table of Contents

About the Guest Editor

Eugene Olevsky is a Distinguished Professor of Mechanical Engineering and Associate Dean for Graduate Studies and Research at the College of Engineering of San Diego State University, California; he is also the Director of the Powder Technology Laboratory at the same University. He has obtained two M.S. degrees in Mechanical Engineering and Applied Mathematics and a Ph.D. degree in Materials Engineering from the Ukraine National Academy of Sciences. Prior to working in the US, Dr. Olevsky carried out studies on sintering at Max-Planck Institute for Metal Research in Stuttgart, Germany, and at the Metallurgy and Materials Department of the Catholic University of Leuven in Belgium. Eugene Olevsky is the author of the internationally recognized *Continuum Theory of Sintering*. He is the author of over 500 scientific publications in the area of sintering research. He has supervised scientific sintering studies of more than 100 post-doctoral, graduate, and undergraduate students. Dr. Olevsky's contributions to research and education have been recognized by multiple awards and honors. He is a Fellow of the American Ceramic Society, a Fellow of the American Society of Mechanical Engineers, and a Humboldt Fellow; he is a Full Member of the International Institute of Science of Sintering. Eugene Olevsky serves as a Chair of the series of International Sintering Conferences—the major forum on sintering science and technology in the world. Prof. Olevsky's most recent research is focused on field-assisted sintering techniques and sintering-assisted additive manufacturing.

Preface to "Spark-Plasma Sintering and Related Field-Assisted Powder Consolidation Technologies"

Electromagnetic field-assisted sintering techniques have attracted increasing attention from scientists and technologists. Spark plasma sintering (SPS) and other field-assisted powder consolidation approaches provide remarkable capabilities to the processing of materials into previously unattainable configurations. Of particular significance is the possibility of using very fast heating rates, which, coupled with the field-assisted mass transport, stand behind the purported ability to achieve high densities during consolidation and to maintain the nanostructure of consolidated materials via these techniques. Potentially, SPS and related technologies have many significant advantages over the conventional powder processing methods, including the lower process temperature, the shorter holding time, dramatically improved properties of sintered products, low manufacturing costs, and environmental friendliness.

In this collection, modern trends of field-assisted sintering, including the processing fundamentals and optimization of final product properties, are highlighted and discussed.

The book includes four articles on spark plasma sintering of metallic or metal-ceramic powder systems, six papers on spark plasma sintering of ceramic materials, two papers on the flash sintering effect, two papers on microwave sintering, one paper on high voltage electric discharge compaction, and one paper on electric pulse sintering of powder paste.

Eugene A. Olevsky
Guest Editor

materials

Article

Formation of Aluminum Particles with Shell Morphology during Pressureless Spark Plasma Sintering of Fe–Al Mixtures: Current-Related or Kirkendall Effect?

Dina V. Dudina [1,2,3,*]**, Boris B. Bokhonov** [3,4] **and Amiya K. Mukherjee** [5]

1   Department of Mechanical Engineering and Technologies, Novosibirsk State Technical University,
    K. Marx Ave. 20, Novosibirsk 630073, Russia
2   Lavrentyev Institute of Hydrodynamics SB RAS, Lavrentyev Ave. 15, Novosibirsk 630090, Russia
3   Institute of Solid State Chemistry and Mechanochemistry SB RAS, Kutateladze str. 18,
    Novosibirsk 630128, Russia; bokhonov@solid.nsc.ru
4   Department of Natural Sciences, Novosibirsk State University, Pirogova str. 2, Novosibirsk 630090, Russia
5   Department of Chemical Engineering and Materials Science, University of California, Davis, 1 Shields Ave.,
    Davis, CA 95616, USA; akmukherjee@ucdavis.edu
*   Correspondence: dina1807@gmail.com; Tel.: +7-383-330-4851

Academic Editor: Eugene A. Olevsky
Received: 25 April 2016; Accepted: 12 May 2016; Published: 14 May 2016

Abstract: A need to deeper understand the influence of electric current on the structure and properties of metallic materials consolidated by Spark Plasma Sintering (SPS) stimulates research on inter-particle interactions, bonding and necking processes in low-pressure or pressureless conditions as favoring technique-specific local effects when electric current passes through the underdeveloped inter-particle contacts. Until now, inter-particle interactions during pressureless SPS have been studied mainly for particles of the same material. In this work, we focused on the interactions between particles of dissimilar materials in mixtures of micrometer-sized Fe and Al powders forming porous compacts during pressureless SPS at 500–650 °C. Due to the chemical interaction between Al and Fe, necks of conventional shape did not form between the dissimilar particles. At the early interaction stages, the Al particles acquired shell morphology. It was shown that this morphology change was not related to the influence of electric current but was due to the Kirkendall effect in the Fe–Al system and particle rearrangement in a porous compact. No experimental evidence of melting or melt ejection during pressureless SPS of the Fe–Al mixtures or Fe and Al powders sintered separately was observed. Porous FeAl-based compacts could be obtained from Fe-40at.%Al mixtures by pressureless SPS at 650 °C.

Keywords: spark plasma sintering; inter-particle; pressureless; iron aluminide; Kirkendall effect

---

1. Introduction

In electric current-assisted sintering of conductive materials, the current passes directly through the compact making inter-particle contacts parts of the electric circuit. The initial resistance of the inter-particle contacts is inherently high due to several reasons, including small diameter of the contact spot and the presence of oxide films and inter-particle gaps. Processes occurring at inter-particle contacts play a key role in the formation of bulk materials from separate powder particles and, thus, require special attention. If pulsed current is applied, the contacts that complete the electric circuit change with every pulse leading to uniform sintering [1]. The contact formation mechanisms between particles of the same material during electric current-assisted sintering have been addressed in detail in a number of studies [2–11]. Burenkov *et al.* [2,3] found evidence of electric erosion between metal particles during electric discharge sintering. Aman *et al.* [6] observed an unconventional morphology of necks formed between copper particles during pressureless Spark Plasma Sintering (SPS) and proposed an ejection mechanism of the inter-particle interactions. Modeling has shown that for fine metallic particles, due to fast heat conduction into the particle volume, no local melting at the contacts should be expected [8]. At the same time, the quality of contacts determines the thermal energy involved in sintering of the compact as a whole. Chaim [11] has recently pointed out that it is correct to discuss the plasma and spark effects for non-conducting materials only, as non-conducting particles can accumulate electric charge. In electrically conducting materials, inter-particle contacts experience excessive Joule heating. Furthermore, inter-particle contacts can be the sites of localized chemical reactions. Vasiliev *et al.* [5] suggested that a high strength of porous zeolite monoliths produced by SPS was due to strong inter-particle bonding established as a result of breakage and rearrangement of chemical bonds in the corresponding areas.

For practical applications of SPS, it is necessary to study the interaction between particles in real systems—multi-component powder mixtures. Certain steps have been made in this direction. Interdiffusion between Ni and Cu particles occurring in three dimensions during SPS was described by Rudinsky and Brochu [12]. Murakami *et al.* [13] studied the formation of compacts from Nb–Al, Nb–Al–W and Nb–Al–W mixtures during SPS with a goal to understand the mechanisms involved in the formation of dense sintered alloys. Kol'chinskii and Raichenko [14] obtained diffusion profiles in the contact region between particles of Ni and Cu formed during electric discharge sintering and laser treatment and found that the diffusion distances were twice as long as those predicted by theoretical calculations without considering the highly localized evolution of heat at the inter-particle contacts. The development of contacts between particles of dissimilar metals can be associated with the formation of solid solutions and intermetallic compounds. Enhanced reaction kinetics in Fe–Al [15] and Mo–Si [16] layered assemblies subjected to treatment in the SPS has been reported. Aiming at dense composite ceramics, Wu *et al.* [17] compared the microstructure uniformity of ceramic composites produced by pressure-assisted reactive SPS and HP and concluded that when compacts with close relative densities were obtained, those sintered by SPS tended to be more homogeneous and of a finer microstructure. This difference was attributed to high heating rates and short holding time in the SPS method. However, the peculiarities of the inter-particle interactions in compacts consolidated from binary mixtures of metals during pressureless SPS have not been investigated. In this work, we present the evolution of particle morphology in Fe–Al mixtures under conditions of pressureless SPS at 500–650 °C. We also make a comparison of the consolidation outcomes achieved by chemical reaction-accompanied pressureless SPS and sintering in a hot press without the application of electric current to the compact.

2. Materials and Methods

For conveniently tracing the morphology changes of the particles in the sintered compacts, Fe and Al powders of spherical morphology were selected. Carbonyl iron (99%, 2.5–5 μm, "SyntezPKZh", Dzerzhinsk, Russia) and gas-atomized aluminum (99.9%, PAD-6, average size 6 μm, "VALKOM-PM", Volgograd, Russia) were used to prepare the Fe-40at.%Al mixtures. Spark Plasma Sintering was carried out using a SPS Labox 1575 apparatus (SINTER LAND, Inc., Nagaoka, Japan). A graphite die of a 10-mm inner diameter and 50-mm outer diameter and short graphite punches of 10 mm diameter were used. A schematic of the assembly used for the pressureless SPS experiments is shown in Figure 1. The die wall was lined with a graphite foil. Circles of graphite foil were placed between the punch and the sample. The temperature during the SPS was controlled by a K-type thermocouple NSF600 (CHINO, Tokyo, Japan) placed in the die wall at a depth of 5 mm. The maximum SPS-temperatures were 500, 600, and 650 °C. The sample was held at the maximum temperature for 3 min and then was cooled down to room temperature. Hot pressing (HP) experiments were conducted with and without applied pressure at 650 °C with a holding time of 5 min. Heating of the sample in the hot press was realized by using external heaters. The applied pressure during HP was 3 MPa. An additional 2 min of holding time were added in HP to ensure the uniform heating of the die. Both SPS and HP were conducted in vacuum. The temperature during HP was controlled by a pyrometer focused on the die wall. The heating rate was 50 °C·min^{-1} in all SPS and HP experiments. Graphite foil was used in HP experiments in a way similar to the SPS experiments. Loose packing of the Fe-40at.%Al powder mixture corresponding to a density of 2 g·cm^{-3} and a relative density of 38% was the initial state of the samples before consolidation, if not stated otherwise. The powder mixture was poured into SPS or HP graphite dies without any additional pressing step. A denser packing with a relative density of 65% was also used in several SPS experiments, which is specified in the specimens' descriptions. Pure Al and pure Fe compacts were obtained starting from loose packing of the corresponding powders. During pressureless SPS and pressureless sintering in the hot press, the only load that the samples experienced was caused by the weight of the upper punch. In pressureless SPS, the die supported a certain pressure applied for electrical contact between the spacers to be established. Annealing of the powder mixture in a tube furnace was conducted at 600 °C for 30 min in a flow of argon.

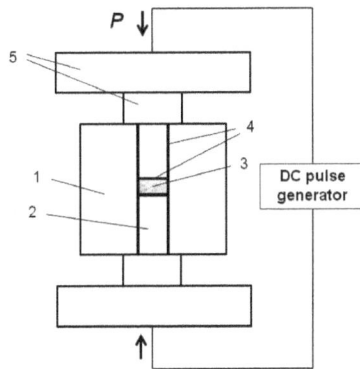

Figure 1. Schematic of the die/punch/spacer assembly used for pressureless Spark Plasma Sintering (SPS) experiments: (**1**) graphite die; (**2**) short graphite punches; (**3**) powder sample; (**4**) graphite foil; (**5**) graphite spacers.

The X-ray diffraction (XRD) patterns were recorded using a D8 ADVANCE diffractometer (Bruker AXS, Karlsruhe, Germany) with Cu Kα radiation. The quantitative phase analysis was conducted using Rietveld analysis of the XRD patterns in PowderCell 2.4 software [18]. The microstructure of the compacts was studied by Scanning Electron Microscopy (SEM) using a Hitachi-Tabletop TM-1000 and a Hitachi-3400S microscope (Hitachi, Tokyo, Japan). The latter is equipped with an Energy-Dispersive Spectroscopy (EDS) unit (NORAN Spectral System 7, Thermo Fisher Scientific Inc., Waltham, MA, USA). Secondary and back-scattered electron ((SE) and (BSE)) images were taken. Selective dissolution treatment of the sintered materials was conducted using 20% NaOH solution at room temperature. The open porosity of the sintered compacts was determined by filling pores with ethanol.

3. Results and Discussion

The ternary carbide AlFe$_3$C was the first phase to form at the inter-particle contacts in the porous compacts. The reflections of AlFe$_3$C can be seen in the XRD pattern of the compact sintered at 500 °C (Figure 2a). In compacts sintered at higher temperatures (Figure 2b–f), the AlFe$_3$C phase was also present as a minor phase. In the absence of carbon, Fe$_2$Al$_5$ was reported to form first upon heating of Fe–Al mixtures [19,20]. There existed a possibility of carbon diffusing from the graphite foil, similar to our previous work, in which Ni$_2$W$_4$C formed during SPS of Ni–W powders [21]. However, in the present study, the main source of carbon was the carbonyl iron powder itself, as the product of annealing of the Fe-40at.%Al mixtures in a flow of argon (with no external carbon sources introduced) also contained AlFe$_3$C as a minor phase. The formation of AlFe$_3$C in the products of reaction between a carbonyl iron powder and an aluminum powder during vacuum annealing was also reported in ref. [22]. The presence of the ternary carbide AlFe$_3$C did not alter the phase sequence with increasing temperature reported for the Fe–Al system in the literature. Moreover, unexpectedly, we gained a means to show that surface layers of contacting Fe and Al particles already chemically interact during SPS at 500 °C, although this interaction is not accompanied by any noticeable morphological changes.

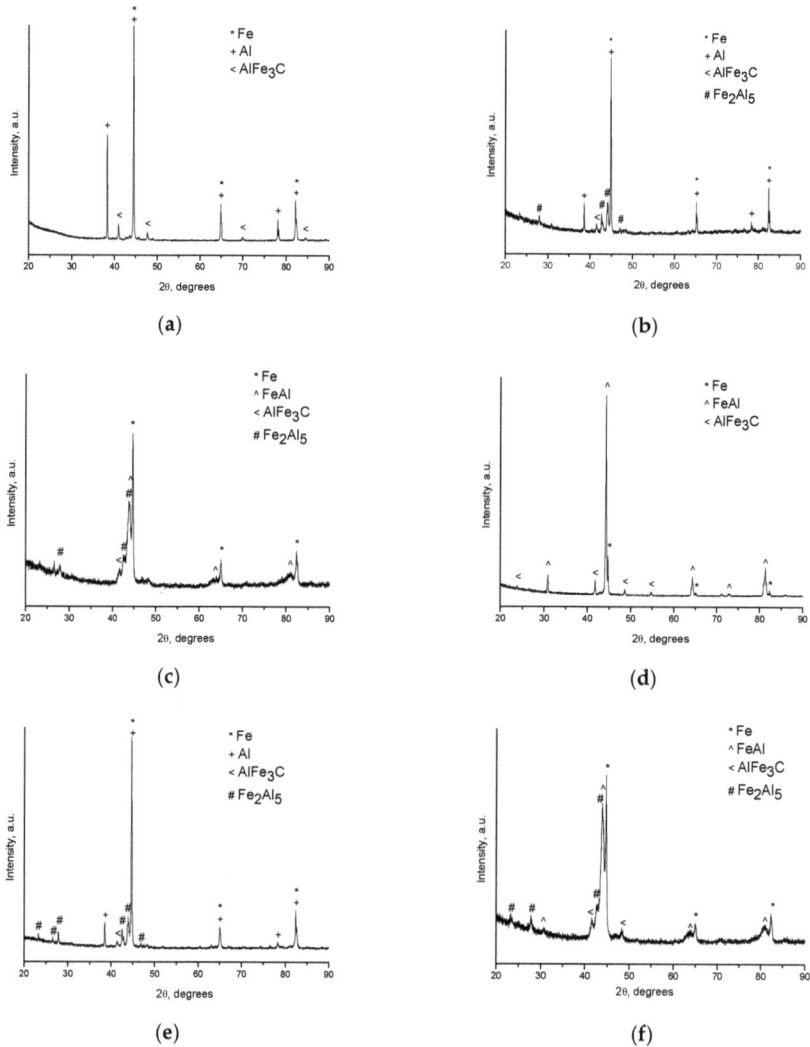

Figure 2. XRD patterns of the porous compacts obtained from Fe-40at.%Al mixtures by pressureless SPS at (**a**) 500 °C (green density 65%); (**b**) 600 °C; (**c**) 600 °C (green density 65%); (**d**) 650 °C and by the hot pressing technique at (**e**) 650 °C, pressureless experiment; (**f**) 650 °C, applied pressure 3 MPa.

As was pointed out by Japka [23], the skin layer of a carbonyl iron particle etches differently compared with the rest of the particle, which is an indirect evidence of structural and chemical differences between the skin layer and the particle volume. Indeed, considering the production process of carbonyl iron powders, the concentration of carbon in the surface layer of particles can be higher than the volume-averaged value.

From the fracture surface of the porous compacts, we can trace the evolution of the inter-particle contacts with temperature and observe the influence of green density and consolidation method of the powder (Figure 3). The spherical morphology of iron and aluminum particles in the compact sintered by SPS at 500 °C starting from a green density of 65% (Figure 3a) is largely maintained. Indeed, intense reflections of the initial components—Al and Fe—can be seen on the corresponding XRD pattern (Figure 2a). SEM did not reveal any evidence of local melting or erosion/melt ejection processes during SPS (Figure 3a). The preserved particle morphology in compacts sintered from loosely packed powders of Al (Figure 4a) and Fe (Figure 4b) powders separately at a temperature of 600 °C confirmed the absence of local melting effects, although both factors—loose initial packing and a higher temperature—could have favored non-conventional inter-particle interactions under applied current. These observations agree with modeling results of ref. [8], which showed that metallic particles several micrometers in diameter cannot sustain the locality of overheating in the inter-particle regions because of high thermal conductivity. In compacts sintered at 600 and 650 °C, because of reaction advancement, it was not possible to define the neck regions in the reaction-sintered porous compacts (Figure 3b–d), as it is usually done in compacts obtained from single-phase powders. The Fe_2Al_5 phase formed in the compacts processed by SPS at 600 °C starting from loose packing, although the initial reactants were still present (Figure 2b). A higher green density of the Fe-40at.%Al mixture resulted in higher transformation degrees of the reactants at the same sintering temperature (Figure 2c). This should be attributed to an increased number of the reaction initiation sites.

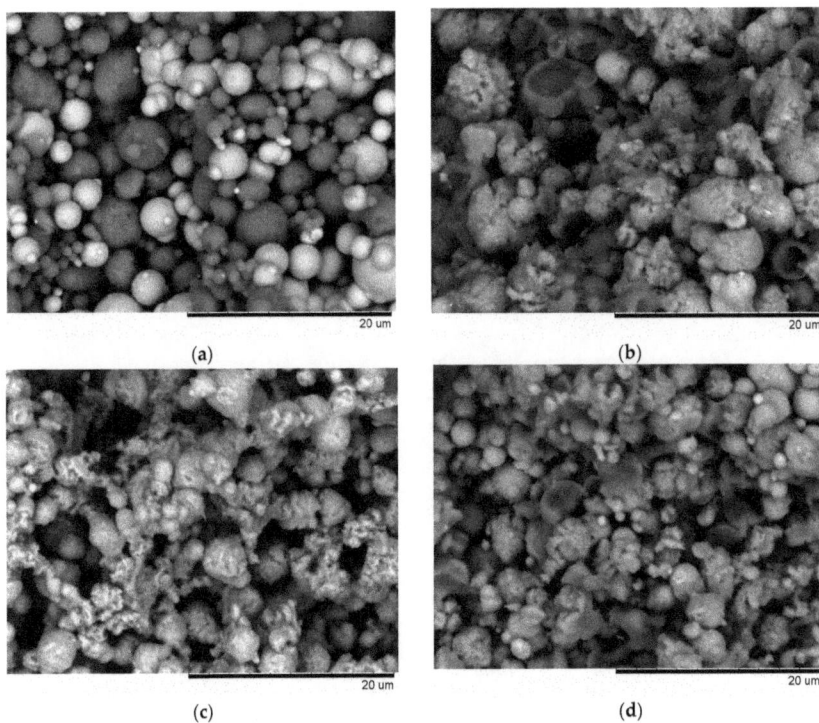

Figure 3. Fracture surface of porous compacts (BSE images) obtained from Fe-40at.%Al mixtures by pressureless SPS (**a**) 500 °C (green density 65%); (**b**) 600 °C; (**c**) 650 °C; (**d**) sintered in a pressureless experiment in the hot press at 650 °C.

Figure 4. Fracture surface of porous aluminum (**a**) and porous iron (**b**) obtained by pressureless SPS at 600 °C.

An interesting observation made in the present study was the formation of particles with shell morphology in the compacts produced by SPS experiencing early chemical interaction stages (Figures 2b and 3b). The observed morphology of seemingly "broken" shells was not due to fracturing of intact hollow particles (that could have been present in the as-sintered sample) during the preparation of samples for SEM observations, as edges of the shells showed a variety of orientations relative to the fracture surface. These shells did not show any specific orientation relative to the current direction during SPS and were also observed on the flat ends of the disk-shaped compacts (Figure 5). The flat ends of the compact were totally free from the graphite foil residue (no sticking occurred) and, therefore, did not require any manipulations to prepare a SEM sample. In a study by Rufino *et al.* [24], a fraction of the initially spherical Al particles showed cavities after heating in argon up to 700 °C, and a reasonable explanation for that was shrinkage upon solidification of the aluminum melt. In those experiments, the cavities had quite smooth edges unlike those of shells formed in the present study (Figure 6a). The EDS mapping (Figure 7) confirms that these shells are partially reacted Al particles. Some Al particles observed on the fracture surface and flat ends of the compacts are of a shape of an apple bitten from different sides. It should be noted that mechanical integrity of the contacts between particles is not maintained in the compacts sintered without the application of pressure from loose packing.

Figure 5. BSE image of the flat end of the disk-shaped compact Spark Plasma Sintered at 600 °C under pressureless conditions from a loosely packed Fe-40at.%Al mixture.

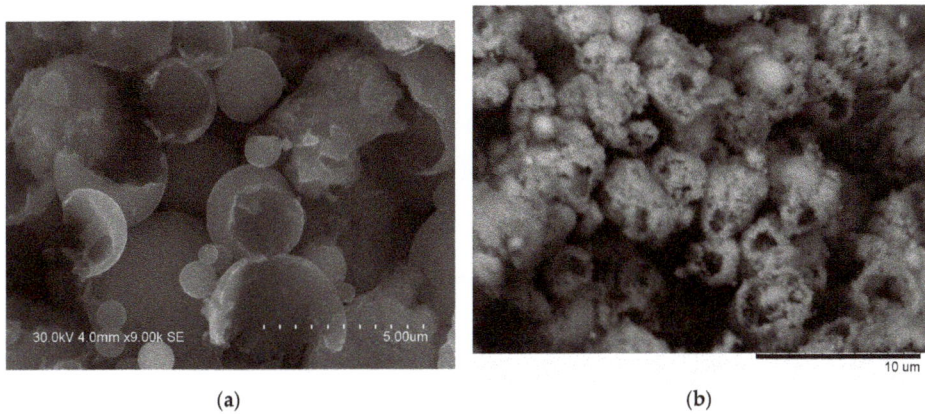

Figure 6. Morphology of the Al shells observed in the compacts formed by Fe-40at.%Al mixtures at an early stage of chemical interaction (**a**) and microstructure of these compacts after treatment in 20% NaOH solution (**b**); (**a**) SE image; (**b**) BSE image.

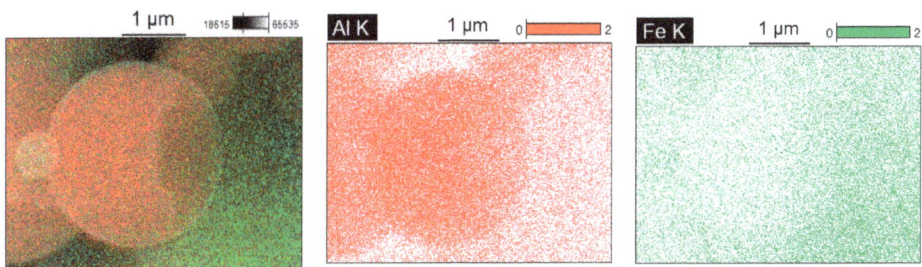

Figure 7. EDS mapping of particles with shell morphology observed in the compacts formed by Fe-40at.%Al mixtures at an early stage of chemical interaction.

It was rather intriguing to look into the origin of the shell morphology. As comparative HP experiments have shown, particles with shell morphology also formed in the compacts consolidated without electric current (Figure 3d). The similarity of the compacts obtained by SPS and HP and showing particles with shell morphology was the early interaction stage with free aluminum still present (Figure 2b,e). As was reviewed by Anderson and Tracy [24], the synthesis of hollow particles and porous materials based on the Kirkendall effect has been conducted in a variety of systems. In the Fe–Al system, the Kirkendall pores form in places of Al particles, as Al rapidly diffuses into Fe and participates in the formation of intermetallic phases. Therefore, it was concluded that the shape of the Al particles observed in this study was due to preferential diffusion of Al into Fe and a further loss in mechanical integrity of the contact between the Al and Fe particles. As Al shells were found in the compacts produced by both SPS and HP, this morphology change was not caused by specific electric current-related effects.

Based on the data presented in Reference [25] on the thickness of the product layers grown at the interface between Fe and Al plates during SPS at 600 °C and a pressure of 5 MPa, we calculated the thickness of the Al layer consumed in the reaction in these conditions. For a holding time at the maximum temperature during SPS of 3 min, the thickness of the Al layer consumed in a planar configuration of the interface is 13 μm. Considering the diameter of the Al particles used in the present work, it may appear that the particles should have been fully consumed. However, in the configuration of Reference [25], the diffusion flow occurred in a single direction normal to the interface between the plates. In the present work, Al diffused into contacting Fe particles in several directions and in three dimensions. Furthermore, a loss in mechanical integrity of the inter-particle contacts can disrupt the diffusion flows. Treatment in NaOH solution allowed revealing another possible contact evolution scenario. The cores of the hollow particles with Fe_2Al_5 shells (Figure 6b) were the unreacted Al and, thus, easily dissolved in alkaline solution. This morphology was possible to form when an Al particle touched several Fe particles in the compact.

The FeAl was the major phase after SPS at 650 °C (Figure 2d), while the reaction has only started by forming a small quantity of Fe_2Al_5 in the compact processed by pressureless HP at this (measured) temperature (Figure 2e). Even applying a pressure during HP did not result in the same transformation degree as was achieved during SPS (Figure 2d,f). Passing electric current through a mixture of powders is used to initiate a combustion reaction for the synthesis of the target products, if the reaction mixture is conductive [26]. In porous compacts, the inter-particle contacts have to carry high current densities [27], which enhance the diffusion kinetics at the interfaces in the case of dissimilar contacts-contacts between the reactants. As the calculated content of free iron in the compact Spark Plasma Sintered at 650 °C from the Fe-40at.%Al mixtures was only 5 vol.%, it can be concluded that reactive SPS offers a very fast synthesis route of porous FeAl-based materials. The open porosity in this compact (Figure 3c) was 42% of the total compact volume.

From a technological perspective, this work has shown that pressureless reactive SPS is a fast synthesis method of porous Fe–Al intermetallics, which are promising high-temperature materials for environmental applications, such as filtration of gases and liquids containing corrosive species. In our experiments, we have also attempted reactive sintering of the Fe–Al mixtures using a SPS die/punch configuration without the upper punch. We found that the absence of direct contact between the compact and the punch causes significant gradients in the Fe–Al compacts, seen both in the phase composition and microstructure. Therefore, in order to ensure the uniformity of the phase composition, microstructure and pore structure of the FeAl porous intermetallic sintered by SPS, direct contacts between the compact and the two punches should be maintained during sintering.

4. Conclusions

The features of interaction between particles of Fe and Al having diameters of several micrometers forming a porous compact during pressureless SPS were studied. The phase evolution of the system with temperature was traced. At early interaction stages, Al particles acquired shell morphology. It was confirmed that the formation of shells was not related to the influence of electric current but was due to the Kirkendall effect in the Fe–Al system and particle rearrangement in a porous compact. No experimental evidence of local melting or erosion/melt ejection processes during SPS was found.

This study has shown that inter-particle interactions between particles of dissimilar materials are more complex than interactions between particles of the same material during SPS in terms of morphology evolution and morphological changes observed during SPS of reacting systems should be carefully studied to separate the effects related to chemical interaction from those caused by passing current, if any.

Acknowledgments: The authors are grateful to Ivan N. Skovorodin for his help in conducting hot pressing, Natalia V. Bulina for recording XRD patterns of the sintered samples, Arina V. Ukhina for her help with selective dissolution experiments and Alexander G. Anisimov for valuable discussions. The SPS Labox 1575 apparatus belongs to equipment of the Center of Collective Use "Mechanics", SB RAS, Novosibirsk.

Author Contributions: D.V.D. designed the study, carried out the experiments and drafted the manuscript. B.B.B. conducted the SEM/EDS analysis and participated in the preparation of the manuscript. A.K.M. participated in the critical discussion of results. All authors read and approved the final manuscript.

Conflicts of Interest: The authors declare no conflict of interest.

Abbreviations

The following abbreviations are used in this manuscript:

SPS	Spark Plasma Sintering
HP	Hot Pressing
XRD	X-ray diffraction
SEM	Scanning Electron Microscopy
SE	Secondary Electron
BSE	Back-Scattered Electron
EDS	Energy-Dispersive Spectroscopy

References

1. Tokita, M. Spark Plasma Sintering (SPS) method, systems and applications. In *Handbook of Advanced Ceramics: Materials, Applications, Processing and Properties*, 2nd ed.; Somiya, S., Ed.; Academic Press/Elsevier: Atlanta, GA, USA, 2013; pp. 1149–1178.
2. Burenkov, G.L.; Raichenko, A.I.; Suraeva, A.M. Dynamics of interparticle reactions in spherical metal powders during electric sintering. *Soviet Powder Metall. Metal Ceram.* **1987**, *26*, 709–712. [CrossRef]
3. Burenkov, G.L.; Raichenko, A.I.; Suraeva, A.M. Macroscopic mechanism of formation of interparticle contact in electric current sintering of powders. *Soviet Powder Metall. Metal Ceram.* **1989**, *28*, 186–191. [CrossRef]
4. Song, X.; Liu, X.; Zhang, J. Neck formation and self-adjusting mechanism of neck growth of conducting powders in Spark Plasma Sintering. *J. Am. Ceram. Soc.* **2006**, *89*, 494–500. [CrossRef]
5. Vasiliev, P.; Akhtar, F.; Grins, J.; Mouzon, J.; Andersson, C.; Hedlund, J.; Bergström, L. Strong hierarchically porous monoliths by pulsed current processing of zeolite powder assemblies. *Appl. Mater. Interfaces* **2010**, *2*, 732–737. [CrossRef] [PubMed]
6. Aman, Y.; Garnier, V.; Djurado, E. Pressure-less spark plasma sintering effect on non-conventional necking process during the initial stage of sintering of copper and alumina. *J. Mater. Sci.* **2012**, *47*, 5766–5773. [CrossRef]
7. Grigoryev, E.G.; Olevsky, E.A. Thermal processes during high voltage electric discharge consolidation of powder materials. *Scr. Mater.* **2012**, *66*, 662–665. [CrossRef]
8. Ye, Y.; Li, X.; Hu, K.; Lai, Y.; Li, Y. The influence of premolding load on the electrical behavior in the initial stage of electric current activated sintering of carbonyl iron powders. *J. Appl. Phys.* **2013**, *113*, 214902. [CrossRef]
9. Bonifacio, C.S.; Holland, T.B.; van Benthem, K. Evidence of surface cleaning during electric field assisted sintering. *Scr. Mater.* **2013**, *69*, 769–772. [CrossRef]
10. Chawake, N.; Pinto, L.D.; Srivastav, A.K.; Akkiraju, K.; Murty, B.S.; Sankar Kottada, R. On Joule heating during spark plasma sintering of metal powders. *Scr. Mater.* **2014**, *93*, 52–55. [CrossRef]
11. Chaim, R. Liquid film capillary mechanism for densification of ceramic powders during flash sintering. *Materials* **2016**, *9*, 280–287. [CrossRef]
12. Rudinsky, S.; Brochu, M. Interdiffusion between copper and nickel powders and sintering map development during spark plasma sintering. *Scr. Mater.* **2015**, *100*, 74–77. [CrossRef]
13. Murakami, T.; Kitahara, A.; Koga, Y.; Kawahara, M.; Inui, H.; Yamaguchi, M. Microstructure of Nb–Al powders consolidated by spark plasma sintering process. *Mater. Sci. Eng. A* **1997**, *239–240*, 672–679. [CrossRef]
14. Kol'chinskii, M.Z.; Raichenko, A.I. A model investigation of the sintering of metal powders with intense energy release at inter-particle contacts. *Soviet Powder Metall. Metal Ceram.* **1977**, *16*, 585–588. [CrossRef]
15. Li, R.; Yuan, T.; Liu, X.; Zhou, K. Enhanced atomic diffusion of Fe–Al diffusion couple during spark plasma sintering. *Scr. Mater.* **2016**, *110*, 105–108. [CrossRef]
16. Anselmi-Tamburini, U.; Garay, J.E.; Munir, Z.A. Fundamental investigations on the spark plasma sintering/synthesis process III. Current effect on reactivity. *Mater. Sci. Eng. A* **2005**, *407*, 24–30. [CrossRef]
17. Wu, W.-W.; Zhang, G.-J.; Kan, Y.-M.; Wang, P.-L.; Vanmeensel, K.; Vleugels, J.; Van der Biest, O. Synthesis and microstructural features of ZrB_2-SiC-based composites by reactive spark plasma sintering and reactive hot pressing. *Scr. Mater.* **2007**, *57*, 317–320. [CrossRef]

18. Kraus, W.; Nolze, G. *PowderCell for Windows*; V2.4; Federal Institute for Materials Research and Testing: Berlin, Germany, 2000.
19. Gao, H.; He, Y.; Shen, P.; Zou, J.; Xu, N.; Jiang, Y.; Huang, B.; Liu, C.T. Porous FeAl intermetallics fabricated by elemental powder reactive synthesis. *Intermetallics* **2009**, *17*, 1041–1046. [CrossRef]
20. Shen, P.Z.; Song, M.; He, Y.H.; Gao, H.Y.; Zou, J.; Xu, N.P.; Huang, B.Y.; Liu, C.T. Synthesis and characterization of porous Fe-25 wt % Al alloy with controllable pore structure. *Powder Metall. Metal Ceram.* **2010**, *49*, 183–192. [CrossRef]
21. Bokhonov, B.B.; Ukhina, A.V.; Dudina, D.V.; Anisimov, A.G.; Mali, V.I.; Batraev, I.S. Carbon uptake during Spark Plasma Sintering: Investigation through the analysis of the carbide "footprint" in a Ni–W alloy. *RSC Adv.* **2015**, *5*, 80228–80237. [CrossRef]
22. Rabin, B.H.; Wright, R.N. Process for synthesizing compounds from elemental powders and product. U.S. Patent 5 269 830, 14 December 1993.
23. Japka, J.E. Microstructure and properties of carbonyl iron powder. *JOM* **1988**, *40*, 18–21. [CrossRef]
24. Rufino, B.; Boulc'h, F.; Coulet, M.-V.; Lacroix, G.; Denoyel, R. Influence of particles size on thermal properties of aluminium powder. *Acta Mater.* **2007**, *55*, 2815–2827. [CrossRef]
25. Anderson, B.D.; Tracy, J.B. Nanoparticle conversion chemistry: Kirkendall effect, galvanic exchange, and anion exchange. *Nanoscale* **2014**, *6*, 12195–12216. [CrossRef] [PubMed]
26. Morsi, K.; Mehra, P. Effect of mechanical and electrical activation on the combustion synthesis of Al_3Ti. *J. Mater. Sci.* **2014**, *49*, 5271–5278. [CrossRef]
27. Anselmi-Tamburini, U.; Gennari, S.; Garay, J.E.; Munir, Z.A. Fundamental investigations on the spark plasma sintering/synthesis process II. Modeling of current and temperature distributions. *Mater. Sci. Eng. A* **2005**, *394*, 139–148. [CrossRef]

![materials logo] *materials*

MDPI

Article

The Manufacturing of High Porosity Iron with an Ultra-Fine Microstructure via Free Pressureless Spark Plasma Sintering

Guodong Cui [1,2,*], Xialu Wei [2], Eugene A. Olevsky [2,*], Randall M. German [2] and Junying Chen [1]

1   School of Materials Science and Engineering, Southwest Jiaotong University, Chengdu 610031, China; chenjunying@swjtu.edu.cn
2   College of Engineering, San Diego State University, 5500 Campanile Drive, San Diego, CA 92182, USA; xwei@mail.sdsu.edu (X.W.); randgerman@gmail.com (R.M.G.)
*   Correspondence: gdcui@swjtu.edu.cn (G.C.); eolevsky@mail.sdsu.edu (E.A.O.);
    Tel.: +86-136-8848-1468 (G.C.); +1-619-594-6329 (E.A.O.)

Academic Editor: Dirk Lehmhus
Received: 14 May 2016; Accepted: 17 June 2016; Published: 21 June 2016

Abstract: High porosity (>40 vol %) iron specimens with micro- and nanoscale isotropic pores were fabricated by carrying out free pressureless spark plasma sintering (FPSPS) of submicron hollow Fe–N powders at 750 °C. Ultra-fine porous microstructures are obtained by imposing high heating rates during the preparation process. This specially designed approach not only avoids the extra procedures of adding and removing space holders during the formation of porous structures, but also triggers the continued phase transitions of the Fe–N system at relatively lower processing temperatures. The compressive strength and energy absorption characteristics of the FPSPS processed specimens are examined here to be correspondingly improved as a result of the refined microstructure.

Keywords: porous iron; hollow Fe–N powder; free pressureless spark plasma sintering; compressive strength

---

1. Introduction

Porous metallic materials have attracted considerable attention because of their excellent structural and functional properties [1,2]. For porous materials with a similar level of porosity, smaller pores size can provide a larger specific surface and interfacial areas. Reducing pore size also helps to refine the microstructure and improve the mechanical properties [3]. In the past several decades, various porous metal materials have been developed and produced for the need of industrial applications, such as energy absorption [4,5], weight reduction, energy conservation [6], damping noise reduction [7,8], biomedical implants [9], and energy storage [10,11]. However, applications of porous metallic materials have been limited due to their low mechanical properties and complicated preparation process. In recent years, bulk iron-based porous materials have been considered the most promising porous materials due to their excellent mechanical properties, low cost, and extensive application backgrounds [2,12].

Most bulk porous iron-based materials are produced via casting or sintering processes [1,2]. Casting technologies include adding a blowing agent to the molten metal, freeze casting [13,14], and directional solidification in hydrogen, nitrogen, or argon atmosphere [15,16]. Sintering techniques are often used to fabricate isotropic porous metal materials. The porosity, pore size, and pore distribution can be easily controlled during the sintering process by adding pore-forming agents [1,2]. Commonly employed processes are mixing metal powders and space holders, pre-compaction in conventional powder press, removal of space holders (or pore-forming agents), and sintering [17,18]. These space holders or foaming agents include inorganic salt, organics, and titanium hydride (TiH_2) [19–21].

As a matter of fact, environmentally harmful gases and residues might be released into the matrix during the removal of space holders, and the properties of the obtained final product could be negatively influenced [18]. To keep the impacts of space holder as few as possible, rapid sintering techniques have been used to fabricate metal foam materials from hollow metal particles and fibers, as they are able to achieve required densification level in short periods of time even without using space holders [22,23].

Spark plasma sintering (SPS), as an advanced sintering technology, is frequently used to consolidate various ceramic and metal materials at relatively lower temperatures [24,25]. Recently, this technique has been applied to produce porous materials through both free pressureless and conventional setups with the aid of dissolutions of inorganic salt (such as NaCl) [26,27]. Moreover, due to its rapid heating rate, this technique has also been widely applied in fabricating ultra-fine grained materials [28]. One recent study found that the iron nitride powders can be used to fabricate porous iron alloys with ultra-fine grains by conventional SPS, and that the continued Fe–N phase transition process has an obvious effect on grain refinement and pore formation during the sintering process [29]. This study also confirmed that rapid sintering technology is able to fabricate ultra-fine porous metal pellets using ultra-fine porous metal particles as raw materials.

In this study, submicron-sized hollow Fe–N particles were used to fabricate ultra-fine porous iron specimens with high porosity but good mechanical properties via free pressureless spark plasma sintering (FPSPS) at a maximum sintering temperature of 750 °C. Since the hollow structured Fe–N powder is non-toxic, non-flammable, non-polluting, and chemically stable, the use of this powder as a pore-forming agent can bypass the procedure of adding and removing inorganic or organic space holders. The microstructure, phase composition, compressive properties, and energy absorption capability of the obtained products were evaluated and compared to previous reported data. The FPSPS manufacturing of ultra-fine porous iron is here shown to be simple, manageable, and environmentally friendly.

2. Results and Discussion

The synthesized Fe–N powders consist of uniformly submicron iron nitride particles and these particles are extremely agglomerated (Figure 1a). A few pores on the surface of Fe–N powders can be identified through careful examination. The ε-Fe$_3$N and ζ-Fe$_2$N are the main phase compositions of the Fe–N powder based on the X-ray diffraction pattern (Figure 1b). There are no peaks of iron oxide and iron presenting on the X-ray diffraction pattern, which indicates that all iron oxide powders have been completely reduced and nitrided by ammonia. The TEM investigation gives more details of morphological and structural features of the Fe–N powder. A typical TEM bright field image of agglomerated Fe–N powders is shown in Figure 1c. It can be seen that the Fe–N powder has an irregular geometrical shape and particle size ranging from 300 to 500 nm. Since there are brighter areas in the Fe–N particle, the Fe–N powder is shown to have a porous or hollow structure, as black areas usually indicate a fully dense structure in a TEM image. This porous structure was most likely formed during reduction and nitrodation reactions. A thin layer of nitrides was first generated on the powder surface, and the ammonia kept reacting with the internal substance by penetrating into the powder. Large volumes of gas were released during the reduction process, and these gases were not able to escape to the powder surface within a short period of time. Therefore, residual gas bubbles were trapped in the powder and formed the hollow porous structure.

Figure 1d shows the nitrogen adsorption–desorption isotherm and a Barrett–Joyner–Halenda (BJH) pore size distribution of the Fe–N powders. The isotherm shows significant hysteresis, which also indicates the particular characteristics of the fine structure and strong adsorption of the powder. The strong adsorption observed at P/P_0 close to 1.0 is a result of the accessible large pores in the Fe–N particles. The average pore size in the Fe–N powder was measured to be around 89.8 nm by the BJH method, which is in a good agreement with the above-mentioned TEM results. The maximum BET surface area and the maximum pore volume were 1.598 m^2/g and 0.036 m^3/g, respectively. All of the above-mentioned results support the fact that the Fe–N powder has a hollow porous structure and large specific surface area.

Table 1 summarizes the pre-compaction pressure, density, and porosity of green compacts as well as those of sintered specimens. Most pores were retained in the sintered specimens under the FPSPS conditions, and the porosity of the sintered specimens increased with decreasing pre-compaction pressure. After being sintered, approximately 10%–15% of the pores were eliminated with the volume shrinkage. The volume shrinkage mainly came from the reduction of inter-particle pores as a result of inter-particle neck formation and growth during FPSPS.

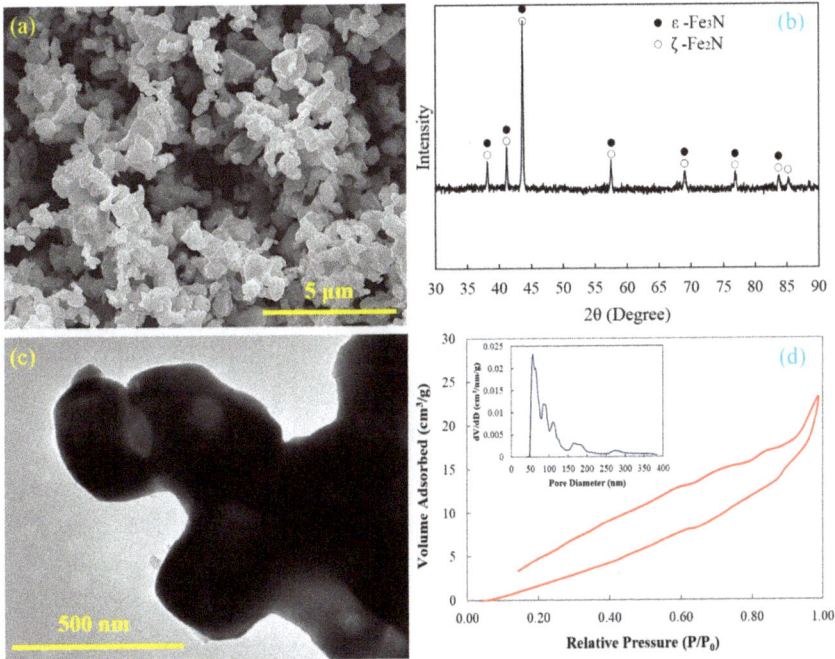

Figure 1. Hollow structure of Fe–N particles: (**a**) SEM image; (**b**) XRD pattern; (**c**) TEM image; (**d**) Adsorption–desorption isotherm and pore size distribution (inset) of Fe–N powder.

Table 1. The pre-compacted pressure, density, and porosity of green compacts and sintered specimens.

Pre-Compacted Pressure, MPa	Green Compact Density, G/Cm^3	Porosity of Green Compacts, %	Sintered Specimen Density, G/Cm^3	Porosity of Sintered Specimens, %
20	2.5	64	3.69	53
40	3.0	57	4.17	47
60	3.2	54	4.40	44

Figure 2 shows the XRD patterns and SEM micrographs of the polished cross section of the sintered specimens. The main composition of the sintered specimen is α-Fe (Figure 2a). As shown in Figure 2b–d, the specimens sintered by FPSPS under different pressures have showed completely different microstructures and pore characteristics. A large number of isotropic pores on a micro- and nanoscale were formed and evenly distributed in the matrix materials, which effectively prevented grain growth and contributed to the finer framework structure.

As one can see, the inter-particle necks were easy to form and grow during the FPSPS process. The increase of pre-compaction pressure contributed to the formation of close-pore structures in sintered specimens (Figure 2b,c). Further, open-pore structures with micro-/nano-pores seemed to be easily observed in the specimens sintered from relatively lower density green compacts (Figure 2d). According to Figures 1b and 2a, and the Fe–N phase transformation process [29], the Fe_2N or Fe_3N can gradually transform into Fe_4N, Fe(N), and Fe as the sintering temperature increases to 750 °C. This transform process also indicates that the nitrogen gas can be produced continually during Fe–N phase transformation. This gas can help to facilitate the formation of pores and prevent grain growing if they are not released in time. Therefore, the porosities in sintered specimens mainly come from inner-particle and the Fe–N phase transformation process, while few come from the inter-particle. In addition, the rapid heating rate, the relatively lower sintering temperature, and the short holding time also contributed to the slow grain growth and facilitated the formation of the ultra-fine porous structure.

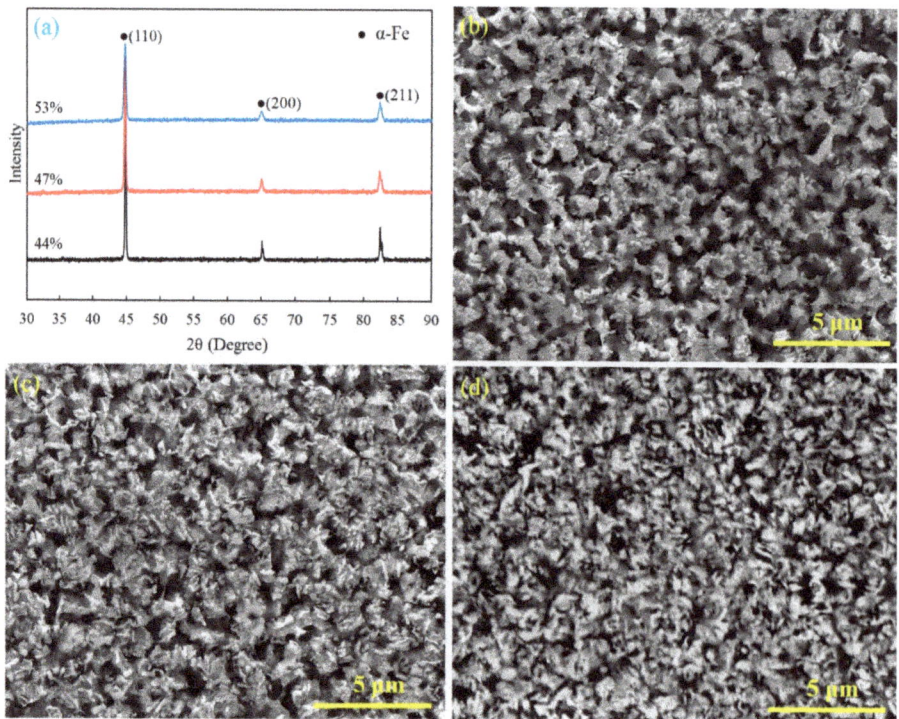

Figure 2. XRD pattern (**a**) and SEM micrographs of porous iron with different porosity: (**b**) 44%; (**c**) 47%; and (**d**) 53%.

The mechanical properties of the ultra-fine microstructure porous iron were examined by uniaxial compressive tests at room temperature. The obtained compressive stress–strain curves are illustrated in Figure 3. These curves have the same evolution tendency and exhibit the typical behavior of ductile porous metal materials [17,18]. The difference is that these curves have not distinguished collapse plateau stage and are only characterized by two regions: In the first linear portion, the compressive stress increases rapidly with increasing strain until the yield point appears at a strain of about 4%. After yield, the compressive stress–strain curves of the porous sintered iron show a gentle ramping up stage, where the stress increases slowly in response to the increase in strain, which indicates a long-term limited deformation strengthening process [30].

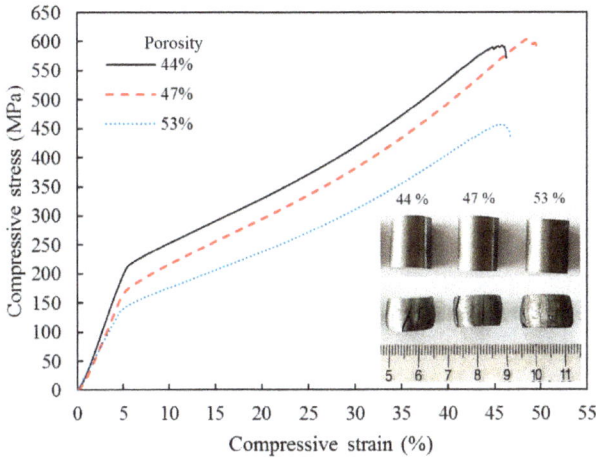

Figure 3. Room temperature uniaxial compressive stress–strain curves of porous iron prepared by pressureless SPS.

The compressive properties and energy absorption properties of sintered specimens are shown in Table 2. It is apparent that either increasing relative density or decreasing porosity corresponds to an increase in Young's modulus and yield strength of the sintered porous iron (Table 2). Young's modulus was measured and calculated from reloading curves after unloading prior to visible plastic deformation. The compressive yield strength was measured as the intercept of tangents taken from the adjacent pre- and post-yield point of the stress–strain curve [17]. The compressive strength is strongly dependent upon the microstructure of the sintered specimens, and the ultra-fine microstructure improves the resistance capability of the porous iron with the bending and the buckling of the "struts". In addition, the Young's modulus of the sintered specimens increased from 3.14 GPa to 4.29 GPa with an increasing density from 3.69 g/cm^3 to 4.40 g/cm^3.

Room temperature compressive properties can be expressed, based on the Gibson–Ashby models utilizing the foam Young's modulus E_f and foam compressive yield strength σ_f, as Equations (1) and (2) [17].

$$E_f = C_E E_S \left(\frac{\rho^*}{\rho_S}\right) m = C_E E_S (1-p)^m \tag{1}$$

$$\sigma_f = C_\sigma \sigma_s \left(\frac{\rho^*}{\rho_s}\right)^k = C_\sigma \sigma_s (1-p)^k \tag{2}$$

where σ_s and E_s are the compressive yield strength and Young's modulus of the bulk material, ρ^*/ρ_s is the relative density of the foam, p is the porosity, C are the scaling factors, and m and k are the constants.

By fitting Equations (1) and (2) with the experimental data (Table 2), the constants in Equations (1) and (2) were optimized to represent the compressive Young's modulus (E_f) and yield strength (σ_f) of the sintered specimens as a function of the relative density (ρ^*/ρ_s) to produce Equations (3) and (4).

$$E_f = 13.5 \left(\frac{\rho^*}{\rho_s}\right)^2 = 13.5(1-p)^2 \tag{3}$$

$$\sigma_f = 1250 \left(\frac{\rho^*}{\rho_s}\right)^3 = 1250(1-p)^3 \tag{4}$$

In Equation (3), the computed results are in a good agreement with the Gibson–Ashby models using a solid modulus E_s of 200 GPa for iron or steel, for which the value of $C_E \approx 0.07$ is found, and

15

the scaling factor (C_E) is lower than the magnitude of the reported values of the scaling factors of the Fe-based foams (0.1–0.3) [17]. This is an indication of a comparatively lower resistance to elastic deflection. From Equation (4), the fitting of the yield strength was not in good agreement with the Gibson–Ashby models. The resulting $C_\sigma \sigma_s = 1250$ MPa, which fitted the experimental data, was much larger than the values of other Fe-based foams ($C_\sigma \sigma_s < 345$ MPa) [17]. This indicates that the strength of the matrix material was greatly improved by refining the microstructure. Such an enhancement is directly related to the grain size, which is smaller in the case of the FPSPS-sintered Fe-based porous materials.

On the contrary, the large numbers of pores uniformly distributed in the iron matrix effectively prevented grain growth and contributed to the formation of a finer framework structure (Figure 2). Thus, the yield strength of the framework was improved remarkably by reducing the grain size. The energy absorption capacity per unit mass (W) and the energy absorption efficiency (η) were calculated from the compressive stress–strain curves (Figure 3) as follows [5]:

$$W = \frac{\int_0^{\varepsilon_m} \sigma \, d\varepsilon}{\rho^*} \tag{5}$$

$$\eta = \frac{\int_0^{\varepsilon_m} \sigma \, d\varepsilon}{\sigma_m \varepsilon_m} \tag{6}$$

where ρ^* is the density of the porous iron, ε_m is the given strain, σ_m is the corresponding compressive stress, σ is the compressive stress as a function of strain ε, and η is the efficiency of the absorbed energy. The absorbed energy per unit mass and the efficiency of energy absorption of the sintered specimens during dynamic compression are shown in Table 2.

Table 2. The compressive properties and energy absorption properties of porous iron sintered by free pressureless SPS.

Porosity %	Young's Modulus, GPa	Yield Strength, MPa	Compressive Strength, MPa	Maximum Strain, %	W kJ/kg	η %
44	4.29	223.1	593.0	45.9	37.20	60.0
47	3.83	178.8	602.0	48.7	39.08	55.6
53	3.14	134.7	456.9	45.8	32.57	57.6

The energy absorption of the porous iron is higher than that of other sintered iron foams with isotropic pores (<30 kJ/kg) [30,31], whereas the energy absorption efficiency of the porous iron is close to 60%. This is mainly because of the higher yield strength and a wider strain range in the long gently stress region (Figure 3). In the dynamic compression of the sintered specimens with 44%, 47%, and 53% porosity at room temperature, the absorbed energy reaches 37.2, 39.08, and 32.57 kJ/kg, respectively.

These energy absorption characteristics of sintered specimens are caused by the two different deformation specifics originating from micro- and nanoscale isotropic pores and matrix metals. In general, for porous metals with isotropic pores, high absorbed energy and high energy absorption efficiency cannot be attained at the same time [30]. Generally, pores are considered defects in solid materials; however, a uniform distribution large number of micro- and nanoscale isotropic pores in the matrix can also have a strengthening effect on the matrix materials by preventing dislocation movement and inhibiting grain growth. These effects are very similar to dispersion strengthening or second-phase strengthening [32]. The energy absorption characteristics can be simultaneously improved along with the matrix strengthening.

3. Materials and Methods

The Fe–N powder utilized in the present study was synthesized using ammonia reduction and nitridation of commercial iron oxide powders (99%, 300 nm, Chengdu Jingke Materials Ltd., Chengdu,

China) at 600 °C for 3 h. The obtained Fe–N powder has an average particle size around 300–500 nm. Weighted Fe–N powders were poured into a 15.3-mm graphite die (I-85 graphite, Electrodes Inc., Santa Fe Spring, CA, USA), whose inner wall had been previously lined with 0.15-mm-thick graphitized paper. Two 15-mm cylindrical graphite punches were used to pre-compact the loaded powder at room temperature within the 15.3-mm die (see Figure 4a). In order to obtain green compacts with different initial densities, the Fe–N powders were pre-compacted under different axial pressures of 20 MPa, 40 MPa, and 60 MPa. After that, these cylindrical graphite punches were removed from the die, and two T-shape graphite punches were placed back to form the free pressureless SPS setup (see Figure 4b). All free pressureless SPS experiments were conducted in a vacuum using a Dr. Sinter SPSS-515 furnace (Fuji Electronic Industrial Co., Ltd., Kawasaki, Japan) [25].

The heating profile is illustrated in Figure 1c: The specimen was first heated up from room temperature to peak temperature at a heating rate of 150°/min and then followed by a 5-min isothermal holding stage in the vacuum (<1 Pa). A 3-kN minimum contact pressure between the die and the T-shape punches was maintained to ensure that the pulsed DC current could go through the tooling components and heat them up rapidly through the Joule heating effect [33]. The maximum processing temperature was selected as 750 °C, as ultra-fine porous structure could be obtained at this temperature according to the Fe–N phase transformation diagram in [29]. The real-time temperature during the SPS process was measured by a K-type thermocouple inserted into a 3-mm depth hole in the middle point of the lateral surface of the graphite die (Figure 4b).

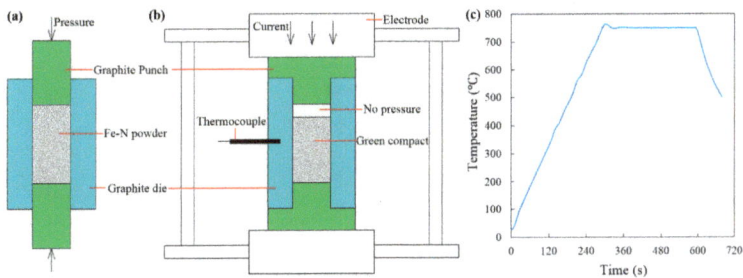

Figure 4. Schematics of the preparation process of porous iron. (**a**) Pre-compaction; (**b**) pressureless SPS process; (**c**) temperature profile used in the pressureless SPS process.

The initial densities of the green compacts were calculated by means of a geometrical method, and the densities of the sintered specimens were measured by means of a water immersion method. The specific surface area (SSA) and pore size distribution of the raw Fe–N powders were determined by nitrogen adsorption–desorption at 77 K using Barrett–Joyner–Halenda (BJH) methods (Quadrasorb. S.I., Quantachrome Instruments, Boynton Beach, FL, USA) after degassing samples at 300 °C for 3 h. The microstructures of Fe–N powders were observed using transmission electron microscopy (TEM, JEM-2100F, JEOL Ltd., Tokyo, Japan) with an accelerating voltage of 200 kV. The microstructures of sintered specimens were observed using scanning electron microscopy (SEM, Quanta 450, FEI Corp., Hillsboro, OR, USA) after etching their cross-sectional areas with 5 vol % Nital. The phase composition of powder and sintered specimens were examined by X-ray diffraction (XRD, X' pert pro, PANalytical B.V., Almelo, The Netherlands) with Cu K-alpha radiation. The Bragg angles were adjusted in the range of 30°–90° for the samples with a scanning rate of 5°/min. The compressive properties of sintered specimens were tested with a uniaxial compression test using a mechanical properties testing system (WDW-200, Changchun Kexin Test Instrument Co., Ltd., Chuangchun, China) with a loading rate of 5 mm/min.

4. Conclusions

In summary, ultra-fine microstructure porous irons with high porosity (>40%) were successfully fabricated by free pressureless SPS at 750 °C using submicron hollow structured Fe–N particles as raw materials. The entire process was environmentally friendly by eliminating the procedures of extra adding and removing space holds. After rapid sintering, a large number of micro- and nano-scaled isotropic pores were formed and evenly distributed in the matrix materials. The continuous Fe–N phase transformation contributed to the formation of the ultra-fine porous structure. The high porosity in the sintered specimens mainly came from the pores in particles, and between particles, and produced during phase transitions in the Fe–N system. These micro- and nano-sized pores and phase transformations in the Fe–N system effectively inhibited grain growth at lower sintering temperatures and markedly refined the microstructure of the matrix materials. The compression stress–strain curves showed a high yield strength and wide strain range with a smooth plateau. Consequently, the energy absorption capability and efficiency were largely improved compared to other metallic foams with isotropic pores.

Acknowledgments: The support of the US Department of Energy, Materials Sciences Division, under Award No. DE-SC0008581 is gratefully acknowledged. The authors from the Southwest Jiaotong University acknowledge the support of Fundamental Research Funds for the Central Universities of China (2682014CX002) and a scholarship from SWJTU, China. The authors would also like to thank Xiaotong Zheng and Jinfang Peng for assistance with the TEM and SEM tests.

Author Contributions: Guodong Cui—literature search, preparation of Fe–N powders, sample preparation, XRD analysis, TEM analysis, manuscript preparation, compression test. Xialu Wei—literature search, manuscript preparation, SEM observations, manufacturing of samples by pressureless SPS. Eugene A. Olevsky —manuscript preparation, results discussion, data analysis, and discussion. Randall M. German—manuscript preparation and results discussion. Junying Chen—data analysis and discussion.

Conflicts of Interest: The authors declare no conflicts of interest.

References

1. Banhart, J. Manufacture, characterisation and application of cellular metals and metal foams. *Prog. Mater. Sci.* **2001**, *46*, 559–632. [CrossRef]
2. Lefebvre, L.-P.; Banhart, J.; Dunand, D.C. Porous metals and metallic foams: Current status and recent developments. *Adv. Eng. Mater.* **2008**, *10*, 775–784. [CrossRef]
3. Tappan, B.C.; Steiner, S.A., III; Luther, E.P. Nanoporous metal foams. *Angew. Chem. Int. Ed.* **2010**, *49*, 4544–4565. [CrossRef] [PubMed]
4. Tane, M.; Zhao, F.; Song, Y.H.; Nakajima, H. Formation mechanism of a plateau stress region during dynamic compression of porous iron: Interaction between oriented cylindrical pores and deformation twins. *Mater. Sci. Eng.* **2014**, *591*, 150–158. [CrossRef]
5. Qiao, J.C.; Xi, Z.P.; Tang, H.P.; Wang, J.Y.; Zhu, J.L. Compressive property and energy absorption of porous sintered fiber metals. *Mater. Trans.* **2008**, *12*, 2919–2921. [CrossRef]
6. Liu, P.S.; Chen, G.F. *Porous Materials: Processing and Applications*, 1st ed.; Elsevier Ltd.: Oxford, UK, 2014; pp. 113–188.
7. Köhl, M.; Bram, M.; Moser, A.; Buchkremer, H.P.; Beck, T.; Stöver, D. Characterization of porous, net-shaped NiTi alloy regarding its damping and energy-absorbing capacity. *Mater. Sci. Eng.* **2011**, *528*, 2454–2462. [CrossRef]
8. Li, Q.Y.; Jiang, G.F.; Dong, J.; Hou, J.W.; He, G. Damping behavior and energy absorption capability of porous magnesium. *J. Alloys Compd.* **2016**, in press. [CrossRef]
9. Wu, S.L.; Liu, X.M.; Hu, T.; Chu, P.K.; Ho, J.P.Y.; Chan, Y.L.; Yeung, K.W.K.; Chu, C.L.; Hung, T.F.; Huo, K.F.; *et al.* Biomimetic hierarchical scaffold: Natural growth of nanotitanates on three-dimensional microporous Ti-based metals. *Nano Lett.* **2008**, *11*, 3803–3808. [CrossRef] [PubMed]
10. Klein, M.P.; Jacobs, B.W.; Ong, M.D.; Fares, S.J.; Robinson, D.B.; Stavila, V.; Wagner, G.J.; Arslan, I. Three-dimensional pore evolution of nanoporous metal particles for energy storage. *J. Am. Chem. Soc.* **2011**, *133*, 9144–9147. [CrossRef] [PubMed]

11. Chen, J.; Yang, D.; Jiang, J.; Ma, A.; Song, D. Research progress of phase change materials (PCMs) embedded with metal foam (a review). *Proc. Mater. Sci.* **2014**, *4*, 369–374. [CrossRef]
12. Murakami, T.; Akagi, T.; Kasai, E. Development of porous iron based material by slag foaming and its Reduction. *Proc. Mater. Sci.* **2014**, *4*, 27–32. [CrossRef]
13. Jung, H.-D.; Yook, S.-W.; Jang, T.-S.; Li, Y.; Kim, H.-E.; Koh, Y.-H. Dynamic freeze casting for the production of porous titanium (Ti) scaffolds. *Mater. Sci. Eng.* **2013**, *33*, 59–63. [CrossRef] [PubMed]
14. Matson, D.M.; Venkatesh, R.; Biederman, S. Expanded polystyrene lost foam casting—Modeling bead steaming operations. *J. Manuf. Sci. Eng.* **2016**, *129*, 425–434. [CrossRef]
15. Nakajima, H. Fabrication, properties and application of porous metals with directional pores. *Prog. Mater. Sci.* **2007**, *52*, 1091–1173. [CrossRef]
16. Kashihara, M.; Hyun, S.K.; Yonetani, H.; Kobi, T.; Nakajima, H. Fabrication of lotus-type porous carbon steel by unidirectional solidification in nitrogen atmosphere. *Scr. Mater.* **2006**, *54*, 509–512. [CrossRef]
17. Scott, J.A.; Dunand, D.C. Processing and mechanical properties of porous Fe–26Cr–1Mo for solid oxide fuel cell interconnects. *Acta Mater.* **2010**, *58*, 6125–6133. [CrossRef]
18. Bakan, H.I. A novel water leaching and sintering process for manufacturing highly porous stainless steel. *Scr. Mater.* **2006**, *55*, 203–206. [CrossRef]
19. Dehaghani, M.T.; Ahmadian, M.; Beni, B.H. Fabrication and characterization of porous Co–Cr–Mo/58S bioglass nano-composite by using NH_4HCO_3 as space-holder. *Mater. Des.* **2015**, *88*, 406–413. [CrossRef]
20. Laptev, A.; Bram, M. Manufacturing hollow titanium parts by powder metallurgy route and space holder technique. *Mater. Lett.* **2015**, *160*, 101–103. [CrossRef]
21. Gu, Y.W.; Yong, M.S.; Tay, B.Y.; Lim, C.S. Synthesis and bioactivity of porous Ti alloy prepared by foaming with TiH_2. *Mater. Sci. Eng.* **2009**, *29*, 1515–1520. [CrossRef]
22. Tang, H.; Wang, J.; Ao, Q.; Zhi, H. Effect of pore structure on performance of porous metal fiber materials. *Rare Met. Mater. Eng.* **2015**, *44*, 1821–1826.
23. Zhou, W.; Tang, Y.; Liu, B.; Son, R.; Jiang, L.; Hui, K.S.; Hui, K.N.; Yao, H. Compressive properties of porous metal fiber sintered sheet produced by solid-state sintering process. *Mater. Des.* **2012**, *35*, 414–418. [CrossRef]
24. Olevsky, E.; Kandukuri, S.; Froyen, L. Consolidation enhancement in spark-plasma sintering: Impact of high heating rates. *J. Appl. Phys.* **2007**, *102*, 114913–114924. [CrossRef]
25. Li, W.; Olevsky, E.A.; McKittrick, J.; Maximenko, A.L.; German, R.M. Densification mechanisms of spark plasma sintering: Multi-step pressure dilatometry. *J. Mater. Sci.* **2012**, *47*, 1–11. [CrossRef]
26. Yamanoglu, R.; Gulsoy, N.; Olevsky, E.A.; Gulsoy, H.O. Production of porous $Ti_5Al_{2.5}Fe$ alloy via pressureless spark plasma sintering. *J. Alloys Compd.* **2016**, *680*, 654–658. [CrossRef]
27. Łazińska, M.; Durejko, T.; Lipiński, S.; Polkowski, W.; Czujko, T.; Varin, R.A. Porous graded FeAl intermetallic foams fabricated by sintering process using NaCl space holders. *Mater. Sci. Eng.* **2015**, *636*, 407–414. [CrossRef]
28. Keller, C.; Tabalaiev, K.; Marnier, G.; Noudem, J.; Sauvage, X.; Hug, E. Influence of spark plasma sintering conditions on the sintering and functional properties of an ultra-fine grained 316L stainless steel obtained from ball-milled powder. *Mater. Sci. Eng.* **2016**, *665*, 125–134. [CrossRef]
29. Cui, G.; Wei, X.; Olevsky, E.A.; German, R.M.; Chen, J. Preparation of high performance bulk Fe–N alloy by spark plasma sintering. *Mater. Des.* **2016**, *90*, 115–121. [CrossRef]
30. Song, Y.H.; Tane, M.; Nakajima, H. Dynamic and quasi-static compression of porous carbon steel S30C and S45C with directional pores. *Mater. Sci. Eng.* **2012**, *534*, 504–513. [CrossRef]
31. Song, Y.H.; Tane, M.; Nakajima, H. Appearance of a plateau stress region during dynamic compressive deformation of porous carbon steel with directional pores. *Scr. Mater.* **2011**, *64*, 797–800. [CrossRef]
32. Bhadeshia, H.K.D.H.; Honeycombe, S.R. *Steels: Microstructure and Properties*, 3rd ed.; Elsevier Ltd.: Oxford, UK, 2006; pp. 17–38.
33. Meng, J.; Loh, N.H.; Tay, B.Y.; Tor, S.B.; Fu, G.; Khor, K.A.; Yu, L. Pressureless spark plasma sintering of alumina micro-channel part produced by micro powder injection molding. *Scr. Mater.* **2011**, *64*, 237–240. [CrossRef]

materials

MDPI

Article

Spark Plasma Co-Sintering of Mechanically Milled Tool Steel and High Speed Steel Powders

Massimo Pellizzari [1,*], Anna Fedrizzi [2] and Mario Zadra [3]

[1] Department of Industrial Engineering, University of Trento, via Sommarive 9, Trento 38123, Italy
[2] Iveco Defence Vehicles, Product Development & Engineering, via Volta 6, Bolzano 39100, Italy; anna.fedrizzi@cnhind.com
[3] K4Sint, via Dante 300–B.I.C., Pergine Valsugana 38057, Italy; mario.zadra@k4sint.com
* Correspondence: massimo.pellizzari@unitn.it; Tel.: +39-0461-282449

Academic Editor: Eugene A. Olevsky
Received: 15 May 2016; Accepted: 9 June 2016; Published: 16 June 2016

Abstract: Hot work tool steel (AISI H13) and high speed steel (AISI M3:2) powders were successfully co-sintered to produce hybrid tool steels that have properties and microstructures that can be modulated for specific applications. To promote co-sintering, which is made difficult by the various densification kinetics of the two steels, the particle sizes and structures were refined by mechanical milling (MM). Near full density samples (>99.5%) showing very fine and homogeneous microstructure were obtained using spark plasma sintering (SPS). The density of the blends (20, 40, 60, 80 wt % H13) was in agreement with the linear rule of mixtures. Their hardness showed a positive deviation, which could be ascribed to the strengthening effect of the secondary particles altering the stress distribution during indentation. A toughening of the M3:2-rich blends could be explained in view of the crack deviation and crack arrest exerted by the H13 particles.

Keywords: mechanical milling; hot work tool steel; high speed steel; spark plasma sintering

---

1. Introduction

Materials for tooling applications, such as tool steels, require a proper compromise between hardness and toughness to provide high wear resistance combined with adequate resistance to cracking. An increased wear resistance often comes at the expense of other properties, such as impact and fracture toughness. The possibility to produce hybrid materials with tailored properties in view of the specific application considered has been proposed as a valuable solution to overcome this problem [1–3].

Powder metallurgy (PM) is a technology that is suited to producing metal matrix composites (MMC). These materials consist of a tough metal matrix that is reinforced by a fine dispersion of hard particles (*i.e.*, carbides, nitrides, borides). To improve the resistance against grooving wear, hard particles must be larger than the abrasive medium [4], but as the hard particle size increases, the tensile and bending strengths of the MMC drastically decrease [5]. To minimize these negative effects, Berns et al. first proposed to reinforce the metal matrix with a harder steel instead of a ceramic material [5,6]. According to Berns' considerations, the present authors also investigated the properties of a hybrid tool steel produced using spark plasma co-sintering (SPS) of a hot worked tool steel (HWTS) and a high speed steel (HSS) [7,8]. SPS is an electric field assisted technology in which a uniaxial pressure combined with a pulsed direct current are applied to produce fully dense materials in a shorter time and at a lower temperature than hot isostatic pressing [9–11].

Previous works showed that it is possible to modulate the hybrid steel properties by changing the composition of the blend [7,8]. However, the co-sintering behaviour of the two steels highlighted a detrimental interaction of the two components and hinders densification [8]. Specifically, the HWTS sinters at a slightly lower temperature than the HSS. The densification of the HSS component is, thus, hindered by the already sintered HWTS skeleton, resulting in the formation of large pores and a considerable decrease in the hardness and toughness. In this respect, a beneficial effect has been demonstrated through the use of small diameter particles, which minimize the interaction between the two components, and the blends achieve nearly full density and good properties [8,12,13].

Mechanical milling (MM) can be successfully used to reduce both the particle size and crystallite size [14–19]. These refinements enhance the sintering process, allowing the production of highly dense materials with better mechanical properties [18,19]. MM can be performed using different technologies [14–16]. In a planetary ball mill, the refinement results from the continuous impacts occurring between the powder particles and balls in the vial. During this high-energy process, the particles are repeatedly flattened, cold welded and fragmented [15]. All of these phenomena are responsible for the morphological and microstructural evolution of the powder [14–17], which can be summarized as follows. In the early stage, the soft and ductile metal particles are easily cold welded by the ball impacts and form large aggregates. This process increases the particle size and considerably changes the particle morphology, which becomes more flat and elongated than in the as-atomized state [15,17]. As the milling process proceeds, the powder particles are continuously strain hardened, becoming progressively less ductile. As a result, their fragmentation by brittle fracture is observed. In this stage, fragmentation prevails over cold welding so that the particle size begins to decrease and the powder shape becomes round again [15,16]. When the particle size becomes too small, the powder particles tend to aggregate again. The system then reaches an equilibrium state in which the agglomerative force and the fragmentation force are balanced. At this optimum stage, the particle size distribution is quite narrow and the mean particle size remains constant at the minimum value [15,16].

In this work, HWTS and HSS powders were mechanically milled to refine their particle size and microstructure. These MM powders were then blended to produce fully dense hybrid steels with different compositions. The blends and two base steels were consolidated using SPS to preserve the fine microstructure obtained during milling [18]. The density, hardness, toughness and microstructure were investigated and compared to those of unmilled blends [8].

2. Materials and Methods

Two commercial gas atomized powders, corresponding to standard grades AISI H13 and AISI M3:2 were used as HWTS and HSS, respectively. Their chemical composition is listed in Table 1.

Table 1. Nominal composition of the powders (wt %).

Material	C	W	Mo	Cr	V	Mn	Si	O	N	Fe
AISI H13	0.41	-	1.60	5.10	1.10	0.35	0.90	0.0105	0.0383	Bal.
AISI M3:2	1.28	6.40	5.00	4.20	3.10	-	-	0.0163	0.0559	Bal.

The starting powders were spherical with 94 wt % of the particles and a diameter less than 350 μm. In both cases, the typical dendritic microstructure produced by rapid solidification could be observed [20].

The route to produce the hybrid tool steel using MM and SPS is schematically represented in Figure 1. The MM was conducted in a Fritsch Pulverisette 6 planetary mono mill at 450 rpm under vacuum. Spheres of 100Cr6 (63HRC) with 10-mm diameters were used, and the ball-to-powder ratio was set to 10:1. To avoid overheating, cycles of 2 min on and 9 min off were used for a total milling time of 1000 min (500 cycles). These parameters were demonstrated to produce an optimum particle size and grain refinement in H13 [20]. The cumulative particle size distribution of the powders was measured using a "Partica LA-950®" (Horiba LTD, Kyoto, Japan) Laser Diffraction/Scattering Particle Size Distribution Analyzer. X-ray diffraction, using both Cu-kα and Mo-kα radiations, was used to identify the phase constitution of base powders and sintered materials.

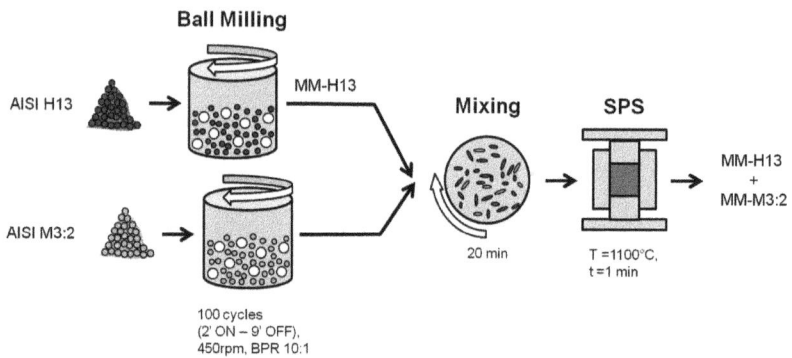

Figure 1. Schematic of the processing route to produce the hybrid tool steel by mechanical milling and SPS.

Conversely, a similar systematic study of the milling conditions for M3:2 was not carried out and this grade was milled using the same parameters for H13. Two different milling runs were carried out for H13 and M3:2. Due to the lack of a suited protection system the powders pick up oxygen and nitrogen when opening the mill vial. The content of these two elements was measured using a LECO TC 400 Analyzer (LECO Corporation, St. Joseph, MI, USA).

Six samples containing different fractions of the two base materials were sintered (Table 2). The blended powders were mixed in a Turbola Mixer for 20 min.

Table 2. Composition and coding of the samples.

Sample Code	Composition (Weight Fraction)	
	AISI H13	**AISI M3:2**
MM-H13	1.0	0.0
MM-80H13	0.8	0.2
MM-60H13	0.6	0.4
MM-40H13	0.4	0.6
MM-20H13	0.2	0.8
MM-M3:2	0.0	1.0

Samples were finally consolidated in a DR. SINTER® SPS1050 apparatus (Sumitomo Coal & Mining Co. Ltd., now SPS Syntex Inc., Tokyo, Japan). Disks with diameters of 30 mm and a 5-mm height were produced in graphite dies. SPS was carried out at 1100 °C with 1 min of isothermal holding at this temperature and final free cooling. The heating rate was 50 °C/min, and a compressive load of 42 kN, which corresponds to a pressure of 60 MPa, was applied once the temperature reached 600 °C. These sintering conditions were selected according to a previous study on the SPS of the as-atomized AISI H13 and AISI M3:2 powders [12]. The holding time was reduced to 1 min only to limit grain growth.

The density was measured using Archimedes' principle according to ASTM B962-08 [21]. The relative density was calculated on the basis of the absolute density of the two MM materials measured using a pycnometer (ρMM-H13 = 7.71 g/cm^3, ρMM-M3:2 = 7.97 g/cm^3). The absolute densities of the four blends were calculated according to the linear rule of mixtures. After standard metallographic preparation and chemical etching, the microstructure of the milled powders and that of sintered materials was observed using scanning electron microscopy (ESEM, Philips model XL30, Philips, Eindhoven, The Netherlands).

All samples were vacuum heat treated by austenitizing at 1050 °C for 15 min and using 5 bar-N_2 gas quenching and double tempering at 625 °C for 2 h each. Hardness was measured using a HV10 scale according to ASTM E92-82 [22]. The apparent fracture toughness, Ka, was determined using a procedure proposed for small fracture toughness specimens [23]. Notch the depth (a) with root radii (ρ) of 50 μm was electro-discharge machined in 6 × 3 × 30 mm³ (W × B × L) specimens. The ratio of the notch depth to the specimen width (a/W) was set at 0.5. Static fracture toughness testing was performed using a 10-ton capacity universal tester. The specimens were loaded in three-point bending at a crosshead speed of 0.5 mm/min according to ASTM E399 [24]. The properties of the current samples were compared with those of samples produced using unmilled powders [8].

3. Results

3.1. Mechanical Milling

A strong particle size refinement is observed in both steels as a result of the MM. The particle size distribution (Figure 2) demonstrates a decrease in the mean size from more than 100 μm (115 μm for H13, 123 mm for M3:2) to less than 20 μm (14.6 for H13, 18.3 for M3:2).

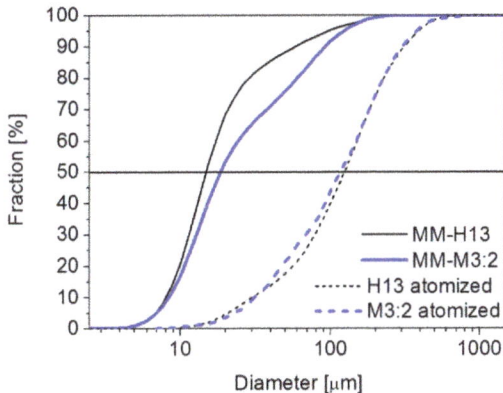

Figure 2. Particle size distribution of base as-atomized and mechanically milled powders.

After only 100 cycles, the two powders showed a round morphology. This morphology did not change during the later stages of the process, as demonstrated by the powders milled for 500 cycles (Figure 3a,b).

A metallographic cross section highlights that some porosity was found inside the particles due to the repeated cold welding and fragmentation phenomena occurring during MM [14–16]. The same effects are also responsible for the disruption of the original inner solidification structure occurring by the stretching and deformation of the dendrites, leading to the formation of a lamellar microstructure [16,17]. As the milling time increased, the lamellae became closer and closer until the microstructure appeared to be fully homogenized. At the end of milling, no more traces of lamellar microstructure could be seen in the AISI H13 (Figure 3c). Conversely, the lower deformation of the MM-M3:2 particles still shows traces of a dendritic structure (see marked regions in Figure 3d), confirming that the MM conditions are far from the optimum and that a greater milling time must be considered to obtain better homogeneity.

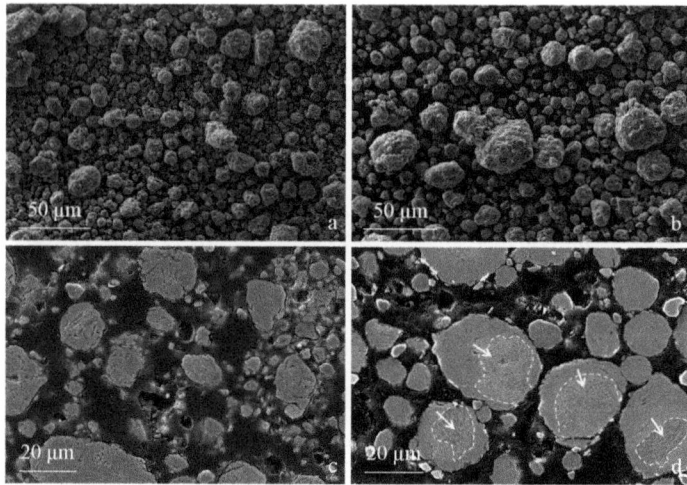

Figure 3. Microstructure (SEM) of the powders (**a**) AISI H13 and (**b**) AISI M3:2 milled for 500 cycles. Metallographic cross section of the same powders (**c**) AISI H13 and (**d**) AISI M3:2 at a greater magnification.

Furthermore, MM was demonstrated to produce a strong structural refinement. A previous study showed that after 500 cycles, the crystallite size, which was measured by X-ray diffraction analysis, decreased from 74 nm to 12 nm in MM-H13 and that, according to the high dislocation density introduced during strain hardening, the hardness increased form 830 HV to 1380 HV [20]. Similarly, the crystallite size of M3:2 decreases from 50 nm to 14 nm in MM-M3:2. Moreover, in both steels, MM promotes the full strain induced transformation of retained austenite (Figure 4). Therefore, MM brings the material to a considerably greater free energy level compared with the original as-atomized state, *i.e.*, more distant from equilibrium, which is a very good starting condition for the faster sintering kinetics of difficult-to-sinter materials, such as those investigated here.

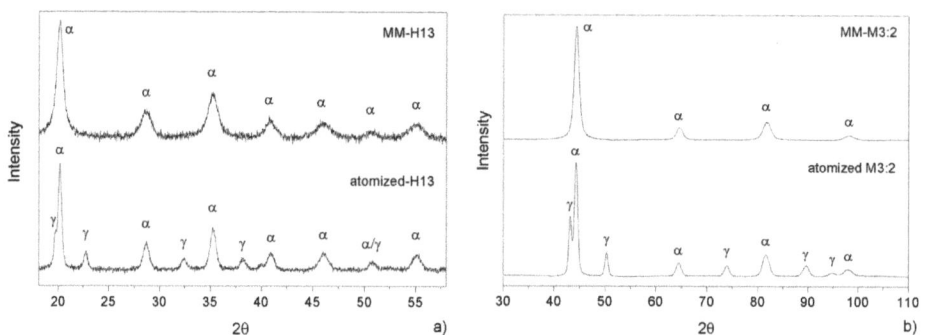

Figure 4. X-ray diffraction patterns for as-atomized and MM (**a**) AISI H13 and (**b**) AISI M3:2 powders.

3.2. Spark Plasma Sintering

3.2.1. Densification

The absolute density of the MM samples linearly decreased as the weight fraction of AISI H13 (*i.e.*, the component with the lower density) increased (Figure 5), which is in good agreement with the linear rule of mixtures. The relative density was calculated as the ratio to the density of the milled powders measured using a pycnometer. Because the milled powders have some internal porosity, especially milled AISI M3:2, these measures are thought to be lower than the theoretical density of the two MM steels. Therefore, the relative density values of the MM materials can be slightly greater than the real values, particularly for specimens with a greater amount of HSS, *i.e.*, MM-M3:2, MM-20H13 and MM-40H13. In any case, the present data confirm that all of the MM samples achieve near full density.

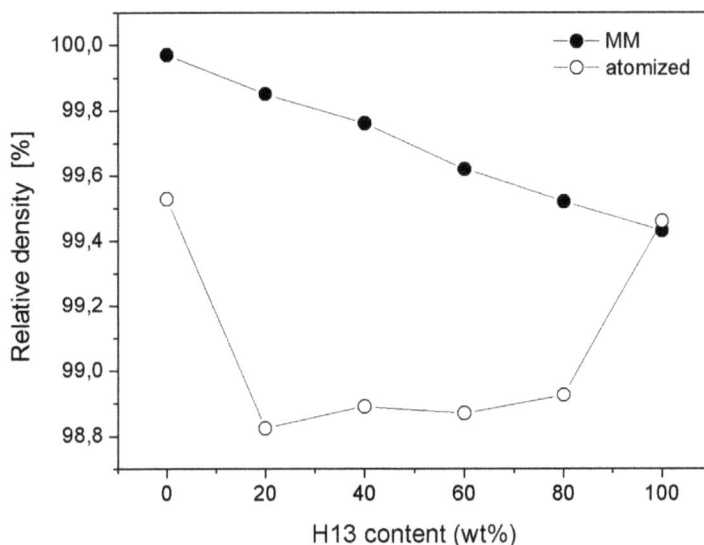

Figure 5. Density and relative density of the sintered specimens as a function of AISI H13 content.

In the case of specimens fabricated using as-atomized powders, the relative density of all blends was less than the density of the two base steels, and all of these blends did not achieve the theoretical density predicted by the linear rule of mixtures. These specimens presented a large amount of porosity, which could be attributed to the different sintering kinetics of the two steels [8]. The sintering process of the as-atomized AISI H13 began at a lower temperature than in AISI M3:2, so the subsequent densification of the HSS was hindered by the presence of a rigid AISI H13 skeleton [8]. The present results confirm that this interaction is significantly minimized after reducing the particle size by MM. Indeed, small AISI H13 particles are less detrimental for achieving a high density because they exert a lower constraint on the AISI M3:2 sintering.

A confirmation can be found in the graph displaying the first derivative of the punch displacement during SPS as a function of temperature (Figure 6).

For a more detailed explanation of the form of this curve, the reader should see the authors' previous papers [8,12,13]. For the purpose of the present discussion, it should be noted that this curve can be representative of the densification rate. When comparing the curves of the MM and as-atomized steels, the MM steel curves are observed to be shifted to a lower temperature, meaning that densification is activated by the milling process. The densification rate of the MM steels is very high, even at a low temperature (700 °C), where the rate of the atomized samples is practically zero. Furthermore, whereas the densification rate abruptly drops at 1050 °C, which is where the densification process of the MM steels is practically concluded, the same does not occur for the atomized steels. Finally, the curves of the MM steels are much closer, which is synonymous with having similar densification kinetics.

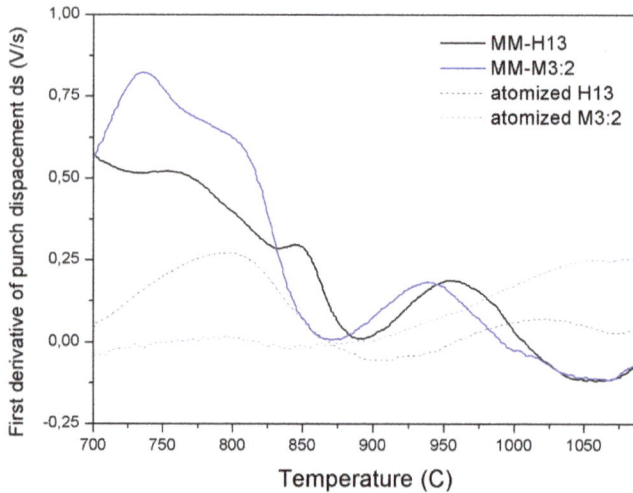

Figure 6. The lower punch displacement during the SPS of MM and the as-atomized base steels.

3.2.2. Microstructure

After SPS the particles of the two steels are very well dispersed and the microstructures of the blends are quite homogeneous (Figure 7). The reduction in the particle size, especially of the largest particles, results in a more uniform microstructure than that produced using unmilled powders. Furthermore, in agreement with the density data, the MM-blends do not show any appreciable porosity which is instead very evident in the as atomized blends (Figure 5) [8].

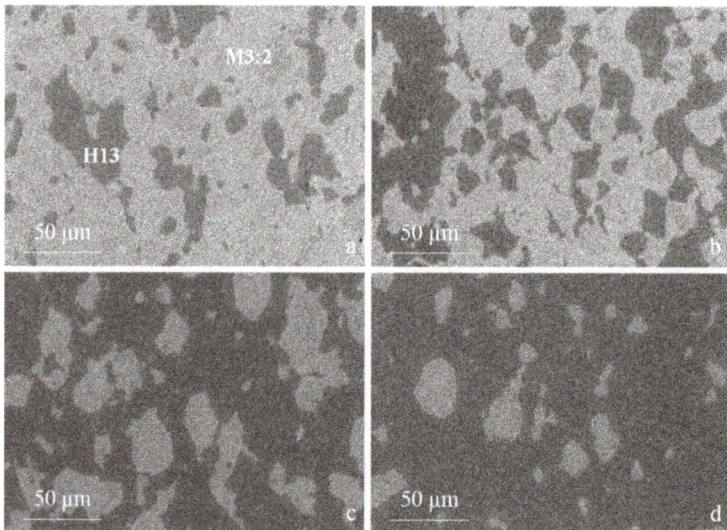

Figure 7. Microstructure of hybrid tool steels (SEM-BSE): MM-20H13 (**a**); MM-40H13 (**b**); MM-60H13 (**c**) and MM-80H13 (**d**). Light and dark regions correspond to M3:2 and H13, respectively.

A closer look at the microstructure of the two base steels shows a much finer grain size compared with as-atomized samples [20]. The grain size indicates that recrystallization occurred during sintering but also that that grain growth could be limited to obtain an average grain size of 0.94 µm and 0.75 µm for H13 (Figure 8a) and M3:2 (Figure 8b), respectively. The present result confirms the suitability of SPS as an evaluable method for the consolidation of nanostructured powders. A quite impressive result is the very fine and homogeneous dispersion of MC (grey particles) and M6C (white particles) carbides in MM-M3:2, which are very effective in ensuring a very small and uniform grain size after sintering. Comparing the XRD patterns in Figures 4b and 9 it can be inferred that these carbides precipitate during sintering. The positive influence of MM, in this respect, can be inferred from the microstructure of the larger HSS particles, which showed a less deformed inner part compared with the smaller particles. After sintering, the microstructure in the core region (1 in Figure 10) demonstrates a less intense carbide precipitation than in the outer shell (2 in Figure 10), resulting in a coarser grain size.

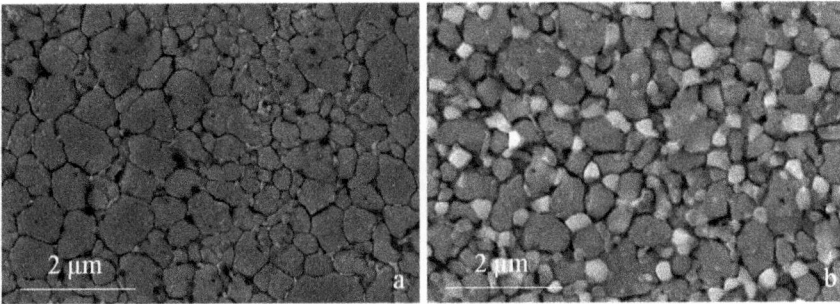

Figure 8. Microstructure of the two base MM steels (**a**) H13; (**b**) M3:2.

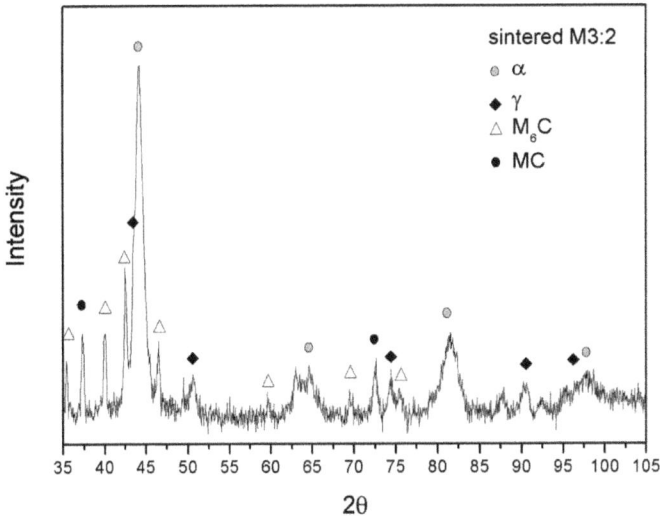

Figure 9. X-ray diffraction pattern for sintered M3:2.

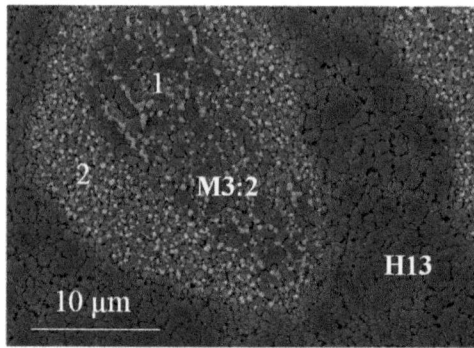

Figure 10. Microstructure of the hybrid tool steel showing the H13 and M3:2 regions. Please note the different grain sizes and carbide distributions in the outer (2) and inner (1) M3:2 particle regions due to the different extent of plastic deformation after milling (see also Figure 3d).

In co-sintered steels, this is reflected in a further interesting result, which is the refinement of the grain size of the H13 neighbouring the HSS particle, *i.e.*, showing the smallest grain size. In other words, the grain growth in H13 is constrained by the grain boundaries of M3:2.

3.2.3. Hardness

The hardness of the MM samples was measured in the as-sintered and the heat-treated state (Figure 11). The high values in the as-sintered state are representative of the primary martensite microstructure forming during the post-SPS cooling stage. As discussed previously, the sintered samples still show the effects of MM, which are reflected in a greater hardness than the as-atomized ones. The temperature-time combination used for sintering plausibly preserves part of the straining effect previously induced by MM so that the hardness of the as-sintered samples (892 HV for M3:2 and 693 HV for H13) is approximately 75 HV greater than that of as-atomized samples (817 HV for M3:2 and 62 7 HV for H13). Further investigation is needed to distinguish any possible influence of the finer grain size from that of dislocations.

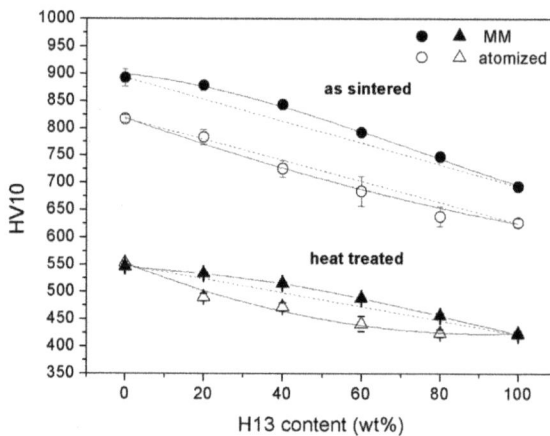

Figure 11. Hardness of the as-sintered and heat-treated hybrid tool steels.

The hardness becomes considerably less after quenching and tempering at 600 °C to obtain the required balance between the hardness and toughness. Tempering above the secondary hardness peak was shown to substantially hide the structural modifications induced by the MM so that the atomized and MM samples of the two base steels show the same hardness [25].

As expected, a lower hardness was observed in AISI H13 due to the lower content of carbon and alloying elements (Table 1), which results in the formation of a softer martensite and a lower amount of carbides. The four MM blends achieve greater hardness values than those predicted by the linear rule of mixtures (dashed line in Figure 11). Previous investigations demonstrated that the dispersion of hard particles in the metal matrix increases the flow resistance and improves the hardness [26]. Furthermore, the investigation of the correlation between the hardness and the tensile strength highlighted that the slight improvement of the tensile strength resulting from the addition of hard particles may correspond to a comparatively greater increase in the hardness [27]. The greater work hardening of the MMCs could be traced back to the local compression of the metallic matrix and the greater concentration of hard particles in the loaded area [27]. The dispersion of hard particles also changes the stress distribution during loading so that stresses greater than the yield stress of the matrix are developed from the initial stage of indentation [26]. Further loading continuously increases the stress and strain in the matrix, causing further work hardening. Consequently, the MMCs show a greater work hardening rate than the metal matrix. For the blends present, the dispersion of the particles of a second constituent cause a similar modification of the stress field, resulting in increased work hardening of the matrix. Conversely, all of the blends produced by the as-atomized powders show a negative deviation from the rule of mixtures highlighting the negative influence of the poor densification [8].

3.2.4. Fracture Toughness

Figure 12 shows the apparent toughness of the MM samples. MM-M3:2 shows less toughness than MM-H13 according to its microstructure and greater hardness. The four MM blends achieve apparent toughness values between those of the two base steels. The values of the two blends containing a greater fraction of AISI H13 (*i.e.*, MM-80H13 and MM-60H13) are close to those predicted by the linear rule of mixtures (dashed line in Figure 12). Although the authors are aware that the fracture toughness of composite materials cannot be simply predicted by the rule of mixtures, the value calculated in this way is reported to highlight the theoretical reference behaviour of a mechanical mix of the two powders.

In MM-40H13 and MM-20H13, the addition of AISI H13 provides a positive deviation from the rule of mixtures, indicating a beneficial influence far beyond that expected by simple mechanical mixing. The reason for the relatively greater toughness of MM-40H13 and MM-20H13 has to be found in the mutual interaction between the two different powders. In the authors' previous experience, AISI H13 produced by SPS generally shows interparticle fractures, resulting in a rough fracture surface [7,8]. Indeed, the same effect can play a toughening role when AISI H13 particles are placed in a less tough matrix. During fracture propagation, the H13 particles force the crack to deviate along their surface (details A in Figure 13a) instead of crossing the M3:2 particles (details B in Figure 13a).

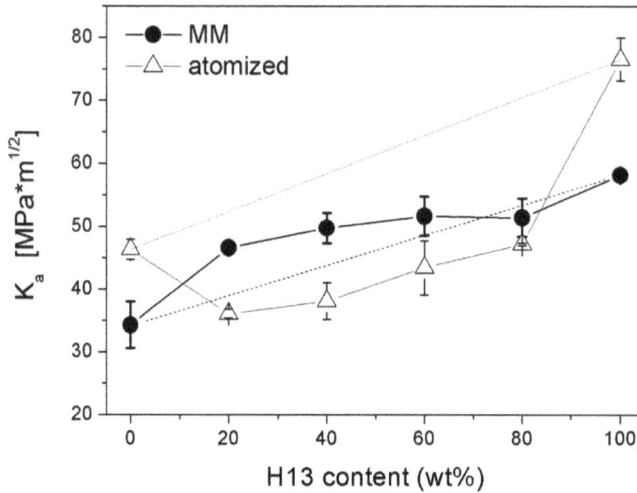

Figure 12. Apparent fracture toughness of tool steels from atomized and mechanically milled powders as a function of the AISI H13 content.

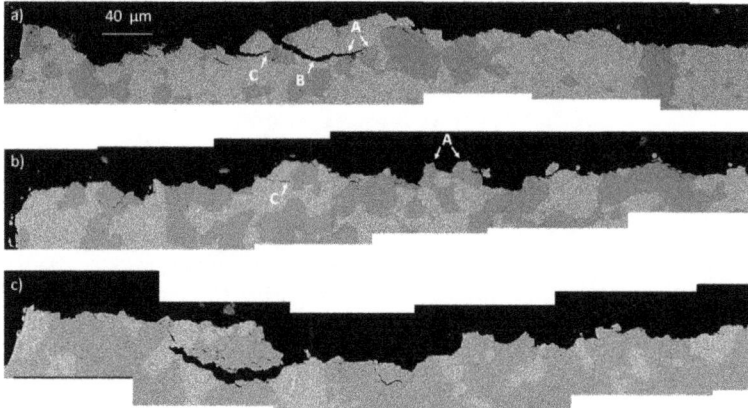

Figure 13. Cross sectional view of the fracture surfaces of (**a**) MM-20H13; (**b**) MM-40H13 and (**c**) MM-60% H13.

This makes the crack path more winding and dissipates more energy, resulting in increased toughness. This explanation is in good agreement with the extent of the observed deviation, which decreases from 20% to 60% H13. By increasing the H13 content, the interparticle spacing progressively decreases, making the crack path less tortuous. In MM-80H13, the HWTS particles are practically interconnected and there is no benefit with respect to the fracture toughness that can be appreciated. Moreover, the H13 particles also act as a barrier against crack propagation (detail C in Figure 13c), suggesting a second possible toughening effect. Conversely, the M3:2 particles do not similarly obstruct the propagation of the crack in the H13-rich blends. Due to the lower toughness, intraparticle cracking is observed in M3:2 such that the crack in H13 proceeds almost straight across them without any toughening effect. Hence the toughness of blends with high H13 fraction decreases according to the linear rule of mixture.

Compared with the as-atomized materials (empty symbols in Figure 12), the apparent toughness of MM-M3:2 decreases from 46 Mpa· $M^{1/2}$ to 34 Mpa· $M^{1/2}$ and that of MM-H13 from 77 Mpa· $M^{1/2}$ to 58 Mpa· $M^{1/2}$. This can be explained in view of the oxygen pick-up shown by the powders after MM (Table 3).

Table 3. Oxygen and nitrogen content in the MM powders.

Material	O (wt %)	N (wt %)
MM-H13 powder	0.1702	0.0978
MM-M3:2 powder	0.1391	0.0950

Due to the lack of a suited insulation system, during post-milling operations (e.g., the delivery of powders to the SPS unit) the contact of highly reactive powders with the environment cannot be avoided, and the oxygen content increased almost one order of magnitude compared with the as-atomized powders (Table 1). The surface of the powders was covered by a thin oxide layer that impairs consolidation during sintering and reduces toughness [7,8,28]. This result suggests that if the oxygen content did not increase, the toughness of all of the MM materials would be 10–20 Mpa· $M^{1/2}$ greater. However, despite of the negative effect of the greater oxygen content, all of the MM blends show greater toughness than the as-atomized blends, which exhibit a greater porosity. It can be concluded that the high porosity due to poor densification is more detrimental for toughness than a high oxygen content. In other words, as far as the present results are concerned, the benefits on densification by MM largely compensate for the detrimental effect of the greater oxygen content. Unquestionably, proper systems aimed at reducing oxidation could bring more benefits than those shown in this research.

4. Conclusions

Hybrid tool steels were successfully produced by mechanical milling and spark plasma sintering of AISI H13 and AISI M3:2 powders. MM markedly reduced the particle size, which minimized the negative influence of the different densification kinetics of the two steels. A high refinement and homogeneous microstructure could be observed for MM-H13, whilst the milling parameters still need to be improved for M3:2. Additionally, near full dense samples (relative density >99.5%) could be sintered for any blend composition.

The results obtained confirm that the properties of the hybrid steel can be modulated by changing the blend composition. The density, hardness and apparent toughness of the blends fall between the values measured for the two base steels according to the H13/M3:2 content. The density values are in good agreement with those predicted by the linear rule of mixtures. Indeed, the hardness of the blends is slightly greater because of the modified stress field distribution in the composite material and the greater local fraction of particles in the plastically deformed steel matrix. An interesting toughening effect by the H13 particles was observed in the MM-M3:2-rich blends. The beneficial effect could be ascribed to two different contributions, namely, the crack deviation and crack arrest exerted by well-dispersed (not interconnected) H13 particles.

The lack of suitable protection against oxidation for MM powders during post-milling operations caused a sharp increase in the oxygen content, resulting in a marked decrease in the toughness for the two base steels. Their toughness is much less than the samples produced using the as-atomized powders. In spite of this, the toughness of the MM-blends is greater than that of the as-atomized blends because the positive influence of a greater density largely compensates for the detrimental influence of the greater oxygen content.

Author Contributions: Massimo Pellizzari and Anna Fedrizzi conceived and designed the experiments; Anna Fedrizzi performed the experiments; Massimo Pellizzari and Anna Fedrizzi analysed the data; Mario Zadra produced the SPS samples and provided a substantial contribution to tune the processing parameters; Massimo Pellizzari wrote the paper.

Conflicts of Interest: The authors declare no conflict of interest.

References

1. Berns, H.; Franco, S.D. Effect of coarse hard particles on high-temperature sliding abrasion of new metal matrix composites. *Wear* **1997**, *203–204*, 608–614. [CrossRef]
2. Pagounis, E.; Lindroos, V.K. Processing and properties of particulate reinforced steel matrix composites. *Mater. Sci. Eng. A* **1998**, *246*, 221–234. [CrossRef]
3. Hannula, S.-P.; Turunen, E.; Koskinen, J.; Soderberg, O. Processing hybrid materials for components with improved life-time. *Curr. Appl. Phys.* **2009**, *9*, 5160–5166. [CrossRef]
4. Zum Gahr, K.-H. Wear by hard particles. *Tribol. Int.* **1998**, *31*, 587–596. [CrossRef]
5. Berns, H.; Von Chuong, N. A new microstructure for PM tooling material. *Met. Phys. Adv. Technol.* **1996**, *6*, 61–71.
6. Berns, H.; Melander, A.; Weichert, D.; Asnafi, N.; Broeckmann, C.; Groß-Weege, A. A new material for cold forging tools. *Comput. Mater. Sci.* **1998**, *11*, 166–180. [CrossRef]
7. Pellizzari, M.; Zadra, M.; Fedrizzi, A. Development of a hybrid tool steel produced by spark plasma sintering. *Mater. Manuf. Process.* **2009**, *24*, 873–878. [CrossRef]
8. Pellizzari, M.; Fedrizzi, A.; Zadra, M. Spark plasma sintering of hot work and high speed steel powders for fabrication of a novel tool steel with composite microstructure. *Powder Technol.* **2011**, *214*, 292–299. [CrossRef]
9. Tokita, M. Development of large-size ceramic/metal bulk FGM fabricated by spark plasma sintering. *Mater. Sci. Forum* **1999**, *308–311*, 83–88. [CrossRef]
10. Munir, Z.A.; Anselmi-Tamburini, U.; Ohyanagi, M. The effect of electric field and pressure on the synthesis and consolidation of materials: A review of the spark plasma sintering method. *J. Mater. Sci.* **2006**, *41*, 763–777. [CrossRef]
11. Olevsky, E.A.; Aleksamdrova, E.V.; Ilyina, A.M.; Dudina, D.V.; Novoselov, A.N.; Pelve, K.Y.; Grigoryev, E.G. Outside mainstream electronic databases: Review of studies conducted in the USSR and post-Soviet countries on electric current-assisted consolidation of powder materials. *Materials* **2013**, *6*, 4375–4440. [CrossRef]
12. Pellizzari, M.; Fedrizzi, A.; Zadra, M. Influence of processing parameters and particle size on the properties of hot work and high speed tool steels by spark plasma sintering. *Mater. Des.* **2011**, *32*, 1796–1805. [CrossRef]
13. Fedrizzi, A.; Pellizzari, M.; Zadra, M. Influence of particle size ratio on densification behavior of AISI H13/AISI M3:2 powder mixture. *Powder Technol.* **2012**, *228*, 435–442. [CrossRef]
14. Benjamin, J.S.; Volin, T.E. The mechanism of mechanical alloying. *Metall. Transact.* **1974**, *5*, 1929–1934. [CrossRef]
15. Maurice, D.; Courtney, T.H. Modeling of mechanical alloying: Part I. Deformation, coalescence and fragmentation mechanism. *Metall. Mater. Trans. A* **1994**, *25*, 147–158. [CrossRef]
16. Suryanarayana, C. Mechanical alloying and milling. *Prog. Mater. Sci.* **2001**, *46*, 1–184. [CrossRef]
17. Çetinkaya, C.; Findik, T.; Özbilen, S. An investigation into the effect of experimental parameters on powder grain size of the mechanically milled 17-4 PH stainless steel powders. *Mater. Des.* **2007**, *28*, 773–782. [CrossRef]
18. Zoz, H.; Ameyama, K.; Umekawa, S.; Ren, H.; Jaramillo, D.V. The millers' tale: High-speed steel made harder by attrition. *Met. Powder Rep.* **2003**, *58*, 18–29.
19. Dai, L.; Liu, Y.; Dong, Z. Size and structure evolution of yttria in ODS ferritic alloy powder during mechanical milling and subsequent annealing. *Powder Technol.* **2012**, *217*, 281–287. [CrossRef]
20. Fedrizzi, A.; Pellizzari, M.; Zadra, M.; Dies, F. Fabrication of Fine Grained Hot Work Tool Steel by Mechanical Milling and Spark Plasma Sintering. In Proceedings of the PM2012 Yokohama, Powder Metallurgy World Congress & Exhibition, Yokohama, Japan, 14–18 October 2012.
21. ASTM International. *Standard Test Methods for Density of Compacted or Sintered Powder Metallurgy (PM) Products Using Archimedes' Principle*; ASTM B962-08; ASTM International: West Coshohocken, PA, USA, 2008.
22. ASTM International. *Standard Test Method for Vickers Hardness of Metallic Materials*; ASTM E92-82; ASTM International: West Coshohocken, PA, USA, 2003.
23. Lee, B.W.; Jang, J.; Kwon, D. Evaluation of fracture toughness using small notched specimens. *Mater. Sci. Eng. A* **2002**, *334*, 207–214. [CrossRef]

24. ASTM International. *Standard Test Method for Plane-Strain Fracture Toughness of Metallic Materials*; ASTM Standard E399-90; ASTM International: West Coshohocken, PA, USA, 1997.

25. Pellizzari, M.; Fedrizzi, A.; Dies, F. Production of a novel hot work tool steel by mechanical milling and spark plasma sintering. In Proceedings of the 9th International Tooling Conference (TOOL2012), Leoben, Austria, 11–14 September 2012; pp. 207–214.

26. Pramanik, A.; Zhang, L.C.; Arsecularatne, J.A. Deformation mechanisms of MMCs under indentation. *Compos. Sci. Technol.* **2008**, *68*, 1304–1312. [CrossRef]

27. Shen, Y.-L.; Chawla, N. On the correlation between hardness and tensile strength in particle reinforced metal matrix composites. *Mater. Sci. Eng. A* **2001**, *297*, 44–47. [CrossRef]

28. Arnberg, L.; Karlsson, A. Influence of powder surface oxidation on some properties of a HIPped martensitic chromium steel. *Int. J. Powder Metall.* **1988**, *24*, 107–112.

materials

MDPI

Article

Bonding of TRIP-Steel/Al₂O₃-(3Y)-TZP Composites and (3Y)-TZP Ceramic by a Spark Plasma Sintering (SPS) Apparatus

Aslan Miriyev [1], Steffen Grützner [2], Lutz Krüger [2], Sergey Kalabukhov [3] and Nachum Frage [3,*]

[1] Department of Mechanical Engineering, Columbia University, New York, NY 10027, USA; aslan.miriyev@columbia.edu

[2] Institute of Materials Engineering, TU Bergakademie Freiberg, Freiberg 09599, Germany; Steffen.Gruetzner@iwt.tu-freiberg.de (S.G.); krueger@ww.tu-freiberg.de (L.K.)

[3] Department of Materials Engineering, Ben-Gurion University of the Negev, P.O. Box 653, Beer Sheva 8410501, Israel; kalabukh@bgu.ac.il

* Correspondence: nfrage@bgu.ac.il; Tel.: +972-8-6461468

Academic Editors: Eugene Olevsky and Dinesh Agrawal
Received: 11 May 2016; Accepted: 7 July 2016; Published: 9 July 2016

Abstract: A combination of the high damage tolerance of TRIP-steel and the extremely low thermal conductivity of partially stabilized zirconia (PSZ) can provide controlled thermal-mechanical properties to sandwich-shaped composite specimens comprising these materials. Sintering the (TRIP-steel-PSZ)/PSZ sandwich in a single step is very difficult due to differences in the sintering temperature and densification kinetics of the composite and the ceramic powders. In the present study, we successfully applied a two-step approach involving separate SPS consolidation of pure (3Y)-TZP and composites containing 20 vol % TRIP-steel, 40 vol % Al₂O₃ and 40 vol % (3Y)-TZP ceramic phase, and subsequent diffusion joining of both sintered components in an SPS apparatus. The microstructure and properties of the sintered and bonded specimens were characterized. No defects at the interface between the TZP and the composite after joining in the 1050–1150 °C temperature range were observed. Only limited grain growth occurred during joining, while crystallite size, hardness, shear strength and the fraction of the monoclinic phase in the TZP ceramic virtually did not change. The slight increase of the TZP layer's fracture toughness with the joining temperature was attributed to the effect of grain size on transformation toughening.

Keywords: fracture toughness; hardness; partially stabilized zirconia (PSZ); shear strength; solid state bonding; phase transformation; spark plasma sintering (SPS); TRIP steel; yttria-stabilized tetragonal zirconia polycrystal ((3Y)-TZP)

1. Introduction

Below 1170 °C, zirconia transforms from the tetragonal phase into a monoclinic structure, accompanied by a volume expansion of 3%–5%. In partially stabilized ZrO₂ (by Y₂O₃, for instance), the tetragonal phase is metastable and displays stress-induced martensitic transformation into the monoclinic structure [1]. In the Nature manuscript named "Ceramic steel?" in 1975, the authors called partially stabilized zirconia "a ceramic analogue of steel" [2]. The toughness of zirconia realized upon such stress-induced transformation was analogously related to that observed in TRIP (transformation-induced plasticity)-steels [2,3]. TRIP-steels exhibit the phase transformation of metastable austenite into α'–martensite during plastic deformation. This phase transformation is called the TRIP-effect and depends on the chemical composition of the steel, temperature and strain rate.

Over recent decades, interest in the fabrication of metal-matrix composites based on TRIP-steels has steadily grown, due to their outstanding properties [4–9]. These steels offer a matrix material presenting a combination of suitable plasticity, high strength and reasonable energy absorption capacity for use as appropriate candidates for high mechanical load applications, such as in structural and safety automotive parts, such as crash absorbers, for instance [4–7]. The unique combination of the TRIP effect in the steel matrix and the transformation toughening of partially stabilized zirconia allows for the creation of composite materials with high damage tolerance [9]. The properties of TRIP-matrix composites with partially stabilized zirconia as a reinforcement were previously addressed [8].

For some applications, such as thermal barriers, the combination of the high damage tolerance of the composites and the extremely low thermal conductivity of pure partially stabilized zirconia may provide sandwich-shaped specimens with controlled thermal-mechanical properties. Among the possible processing techniques for the fabrication of sandwich-shaped specimens from metal and ceramic powders, spark plasma sintering (SPS) has proven to be a suitable approach. SPS is widely used for the consolidation of ceramics [10–15], metals [16–18], intermetallics [19–21] and various composites [22–24]. Accordingly, the properties of composites comprising a high-alloy TRIP-steel reinforced with Al_2O_3, Mg-PSZ or Y-TZP fabricated using SPS have been studied [8,25–31]. However, sintering of a (TRIP-steel-ceramic)/ceramic sandwich in a single step is very difficult due to differences in sintering temperature and the densification kinetics of the composite and the ceramic powders. On the other hand, an SPS apparatus has been successfully employed for the joining of similar and dissimilar materials [32–36]. The remarkable advantages of SPS joining, as compared to hot pressing and hot isostatic pressing, have been previously described [34,35]. Recently, the optimal parameters (i.e., temperature, holding time and applied pressure) for diffusion bonding of various ceramics (e.g., alumina, silicon carbide, boron carbide and magnesium aluminate spinel) using an SPS apparatus were discussed [37]. Yb- and Y-doped α-SiAlON ceramics were effectively diffusion-bonded by an SPS apparatus in less than 20 min at 1650–1700 °C. It was concluded that α-SiAlON grain growth across the joining interfaces modified the joint microstructures so as to secure high bonding strength [38]. The interaction between TRIP-steel and PSZ, particularly the interface phenomena in the TRIP-steel/PSZ system, was investigated using TEM/HRTEM and electron spectroscopy [31]. Grain boundaries between TRIP-steel and PSZ grains of the SPS-processed composite as well as the interface between the PSZ films and TRIP-steel substrate were characterized and compared. It was thus established that the deposited PSZ film was free of cracks and partly coherent with the TRIP substrate. In contrast to the thin film sample, no pronounced heteroepitaxy and/or distinct orientation relationship between TRIP-steel and PSZ grains was observed in the SPS-processed samples. Nevertheless, dislocation clusters and intersecting stacking faults were observed at the grain boundaries.

In the present study, we applied a two-step approach to obtain sandwich-shaped specimens that included separate SPS consolidation of the pure (3Y)-TZP and composites containing 20 vol % TRIP-steel, 40 vol % Al_2O_3 and 40 vol % (3Y)-TZP ceramic phase (hereafter, composites), and the subsequent diffusion joining of both sintered components in the same SPS apparatus. The microstructure and properties of the sintered and bonded specimens were characterized.

2. Results

2.1. Powder Consolidation in SPS

In Figure 1a, the microstructure of the SPS-consolidated (3Y)-TZP specimen is shown. The measured relative density of the (3Y)-TZP samples was 98.4%. Various properties of the (3Y)-TZP ceramic after SPS are shown in Table 1. The microstructure of the sintered composite is shown in Figure 1b. Bright TRIP-steel grains can be seen throughout the ceramic matrix.

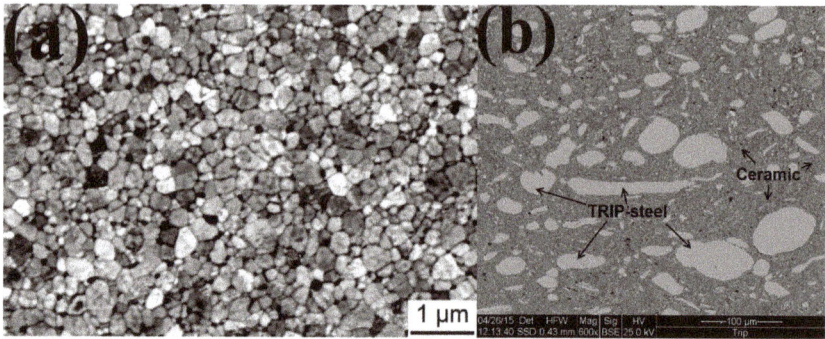

Figure 1. Microstructure of SPS-consolidated specimens: (**a**) (3Y)-TZP; (**b**) composite.

Table 1. Properties of the (3Y)-TZP ceramic and composite.

Property	Value
Grain size (d$_{50}$) of (3Y)-TZP	208 nm
Grain size (d$_{90}$) of (3Y)-TZP	339 nm
Mean crystallite size of (3Y)-TZP	60 nm
Mean crystallite size of the (3Y)-TZP phase in the composite layer	70 nm
Fraction of the monoclinic phase in (3Y)-TZP	<1 vol %
Hardness of (3Y)-TZP	13.2 GPa
Hardness of the composite	10.4 GPa
Indentation fracture toughness (K_{Ic}) of (3Y)-TZP	5.3 MPam$^{0.5}$

2.2. Solid-State Joining in SPS

2.2.1. (3Y)-TZP/(3Y)-TZP Joining

SPS-sintered (3Y)-TZP specimens were self-joined at 1150 °C for 120 min. The HR-SEM image of a slightly etched surface (Figure 2) shows the sub-micron equiaxed zirconia grain structure. The bonding area, marked by a dash-dot red line, is continuous, with no boundary/seam being obtained.

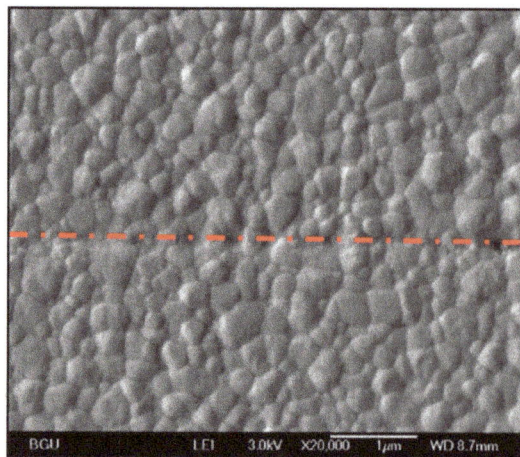

Figure 2. HR-SEM image of a (3Y)-TZP/(3Y)-TZP bonding area cross-section.

Non-destructive analysis of the joint using ultrasonic waves showed no detachments. A graphical representation of the ultrasonic test results can be seen in Figure 3. Each area unit was scanned and the reflection results were translated into a colorimetric/numeric scale, where good bonding is indicated by lower numbers (blue color on the scale) and poor bonding is indicated by higher numbers (red color on the scale). Most of the bonding area, except for some alterations at the sample periphery, showed excellent bonding quality.

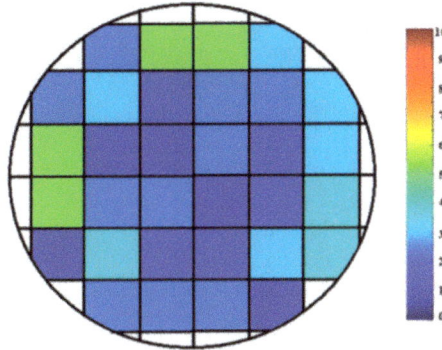

Figure 3. Non-destructive (ultrasonic) analysis of an SPS-joined (3Y)-TZP/(3Y)-TZP specimen. Ultrasonic wave reflections were translated into a colorimetric/numeric/scale reflecting bonding quality (0/blue: high bonding quality, 10/red: low bonding quality).

2.2.2. Composite/(3Y)-TZP Joining

The main problem with the single-step sintering of steel-ceramic/ceramic sandwich structures derives from the different sintering temperatures and kinetics of the metal and ceramic components. Due to the fact that carbon from the graphite die diffuses into the steel, leading to a reduction in the melting point, the sintering temperature of the steel/ceramic composite is limited to 1150 °C. However, significantly higher temperatures (about 1350–1450 °C) are necessary to achieve fully dense (3Y)-TZP. To overcome this problem, solid-state bonding of the sintered composite with a ceramic content of 80 vol % to the sintered (3Y)-TZP part was applied (Figure 4a). No cracks or voids were observed at the bonding interface (Figure 4b).

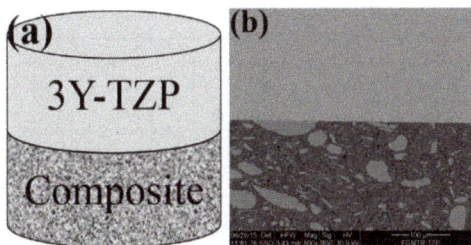

Figure 4. Joining a sintered composite (80 vol % ceramic) to a sintered TZP sample: (**a**) illustration; (**b**) SEM image of the bonding interface (the upper part is the (3Y)-TZP).

A typical microstructure of the (3Y)-TZP layer after joining for 120 min at 1150 °C is shown in Figure 5.

Figure 5. SEM image of a thermally etched surface of the (3Y)-TZP PSZ layer.

Only limited grain growth (from the initial 208 nm to 226 nm after joining at 1150 °C) occurred during joining, while the crystallite size (as revealed by XRD analysis) and the fraction of the monoclinic phase in the TZP ceramic virtually did not change. This means that the martensitic phase transformation from the tetragonal to the monocline structure in the pure TZP did not occur during the joining process. The representative XRD patterns assessed before and after the joining of (3Y)-TZP are shown in Figure 6. The results are in agreement with data reported by Ruiz et al. [39], where significant grain growth and change of phase composition appeared only after isothermal heat treatment at temperatures above 1550 °C.

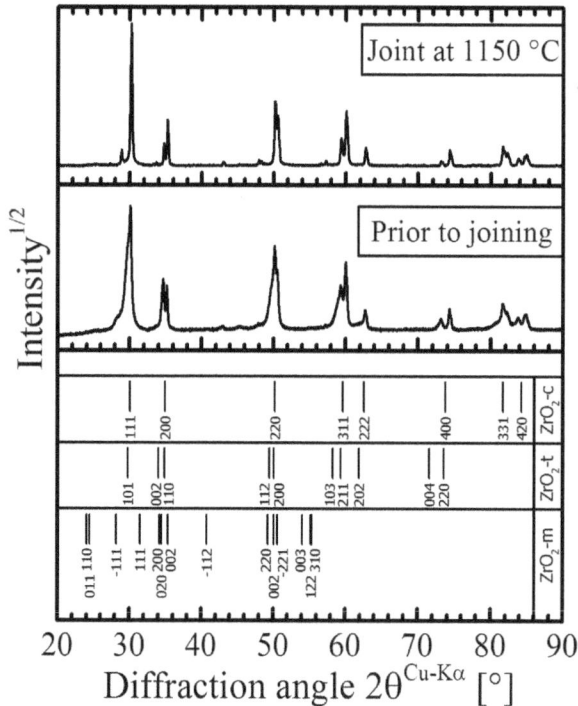

Figure 6. Representative XRD patterns of (3Y)-TZP taken before and after joining at 1150 °C.

The hardness values (Figure 7a) and the shear strengths for broken SPS-bonded samples (Figure 8) likewise did not change during joining, although the fracture toughness of the TZP slightly increased with the increased joining temperature (Figure 7b). This phenomenon may be attributed to the effect of grain size on transformation toughening [40]. Above a critical grain size, tetragonal grains would spontaneously transform into the monocline structure [39–42]. However, the temperature for critical grain growth was not achieved during bonding.

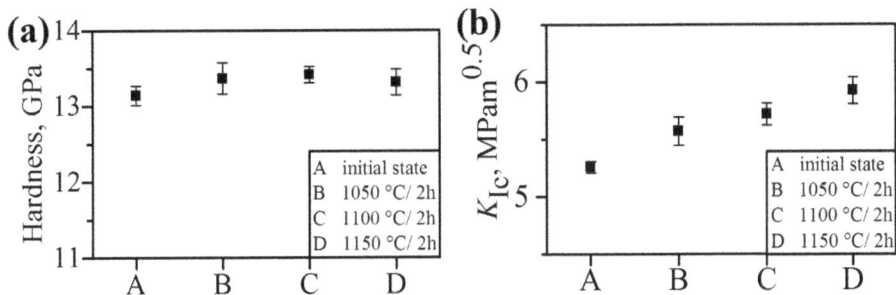

Figure 7. Hardness (**a**) and fracture toughness (**b**) of the TZP layer.

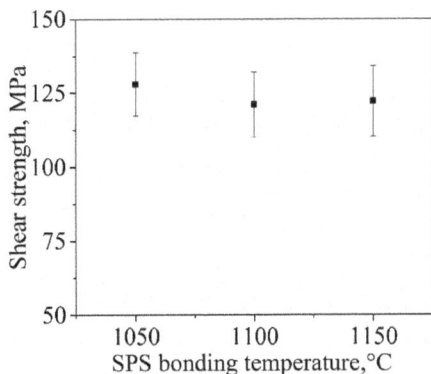

Figure 8. Shear strength of the SPS-bonded specimens plotted against joining temperature.

3. Materials and Methods

3.1. Materials

The composite (20 vol % TRIP-steel, 40 vol % Al_2O_3 and 40 vol % (3Y)-TZP ceramic phase) was synthesized from high alloy austenitic CrMnNi-TRIP-steel (Table 2), alumina and yttria-stabilized zirconia powders. Median particle size of the steel powder was about 17 μm. Alumina (Pengda, Munich, Germany) and (3Y)-TZP ceramic powder (3 mole % yttria-stabilized zirconia, TOSOH, Tokyo, Japan) had median particle sizes of 0.49 μm and 40 nm, respectively.

Table 2. Chemical composition of CrMnNi-TRIP-steel.

Element	Fe	C	Cr	Mn	Ni	Mo	Si	S	N
Wt.%	bal.	0.031	13.50	6.49	5.73	0.07	0.48	0.007	0.034

For composite fabrication, the powders were mixed in a planetary ball mill (Pulverisette 5, Fritsch GmbH, Idar-Oberstein, Germany) under high-energy ball milling conditions for 2 h. The ball material was hardened Cr-steel, the ball diameter was 25 mm, the powder to ball ratio was 1:10 and the rotating speed was 180 rpm.

3.2. Processing

Specimens were fabricated in a two-step process. First, TRIP-steel/Al_2O_3-(3Y)-TZP composite and pure (3Y)-TZP specimens were sintered separately. The powders were consolidated as cylindrical bodies with 20 mm diameter using the SPS technique in FCT-HP D 25/2-2 apparatus. Pure (3Y)-TZP and composite samples had a final height of approximately 3 mm.

The (3Y)-TZP powder was sintered at 1400 °C under uniaxial pressure of 60 MPa for holding time of 5 min. The heating and cooling rates were 100 K/min. The (3Y)-TZP specimens were self-joined in the SPS apparatus at 1150 °C with a holding time of 120 min under an argon atmosphere (10^{-2} torr) and a uniaxial pressure of 16 MPa. The composite was consolidated at 1150 °C for a holding time of 10 min under a pressure of 60 MPa.

The mating surfaces of the samples before joining were prepared by conventional metallographic technique on a 1 μm diamond paste (Struers Nap B) stage, cleaned in acetone and dried in air. The samples were placed in a graphite die with 20 mm inner and 40 mm outer diameters and inserted into the SPS apparatus (FCT–HP D5/1 System, Rauenstein, Germany) for joining. Composite and PSZ specimens were SPS-joined at 1050, 1100 and 1150 °C, respectively, with holding time of 120 min under an argon atmosphere (10^{-2} torr) and a uniaxial pressure of 16 MPa. The temperature was measured by a pyrometer focused on the upper graphite punch. Pulse-mode DC current (pulse 5 ms and pause 2 ms) was used throughout the joining experiments. The cooling rate after bonding was about 12 K/min.

3.3. Characterization

Determination of the density of the samples was achieved using the Archimedes method in distilled water. Quantitative phase analysis was carried out by XRD measurement (Cu-Kα radiation). Phase compositions were estimated using the Rietveld method.

Microstructure was characterized by scanning electron microscopy (SEM). Sample surfaces were ground and polished to a 1 μm diamond finish followed by vibratory polishing for 24 h. After polishing, the prepared ceramic samples were thermally etched at 950 °C for 1 h. Subsequently, grain size and size distribution were determined by the linear intercept method on more than three SEM micrographs, such that over 900 intercepted grains were considered for each condition.

The quality of the (3Y)-TZP/(3Y)-TZP joined region was tested by ultrasonic measurement using a 15 MHz pulse/receiver probe with 3.157 mm diameter (V260, Panametrics, Houston, TX, USA). Joined specimens were divided into squares with an area of about 9 mm^2 (3 mm × 3 mm) each. The specimens were scanned with the transducer and the resulting reflections were documented according to the known sound speed in the tested material and specimen thickness. Reflections from the bonding area and those from the bottom of the tested specimen were collected, classified qualitatively and translated according to a colorimetric scale (blue to red) and numeric scale (ranging from 0 to 10, indicating well and poor bonding quality, respectively).

The maximal shear forces needed for fracture of the composite/(3Y)-TZP bonded specimens were determined using an LRX Plus apparatus (Lloyd Instruments, Fareham Hants, UK). The shear test specimen and tools are shown in Figure 9. Test specimen dimensions were 5 × 5 × 6 mm^3 (width × length × height). The test tools consisted of the static block, into which the specimen was mounted, and the moving block. When mounted into the testing apparatus prior to testing, the static block was placed on the static portion of the apparatus, while its dynamic portion was adjusted to the moving block of the tool. Four specimens were examined for each processing parameter tested.

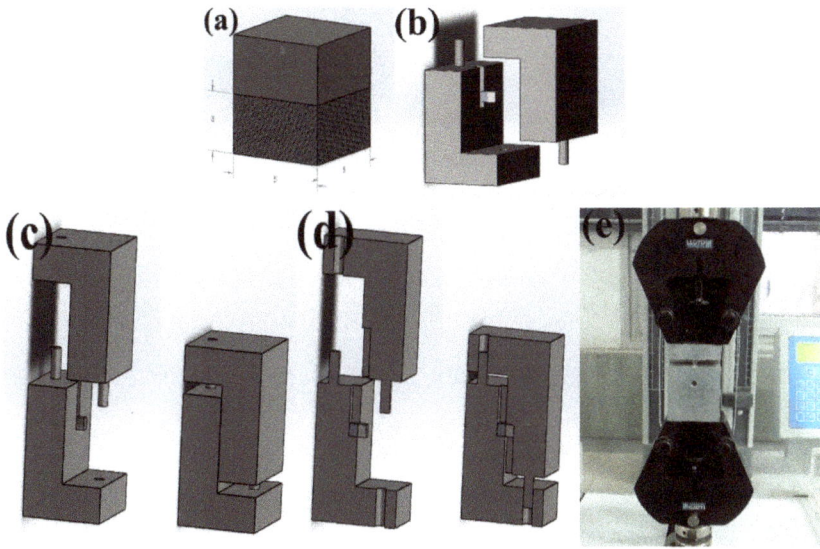

Figure 9. Shear test tools and specimen. (**a**) Test specimen, dimensions $5 \times 5 \times 6$ mm^3; (**b**) test tools and specimen; (**c**) test tools and specimen before and after mounting, full views; (**d**) test tools and specimen before and after mounting, cross-section views; (**e**) testing apparatus with mounted tools.

The fracture toughness of the (3Y)-TZP specimens after sintering and joining was investigated by the indentation method. The applied indentation load P was 98.07 N. Crack lengths were measured using an optical microscope. The fracture toughness K_{Ic} (MPam$^{0.5}$) was calculated from the equation of Niihara et al. [43] for Palmqvist cracks:

$$\left(\frac{K_{Ic} \cdot \varphi}{H_V \cdot \sqrt{a}} \right) * \left(\frac{H_V}{E \cdot \varphi} \right)^{\frac{2}{5}} = 0.035 \left(\frac{l}{a} \right)^{-1/2} \tag{1}$$

where H_V is the Vickers hardness, E is Young's modulus, a represents the half length of the indentation diagonal, l is the mean Palmqvist crack length and φ is a pseudo-constant. The value of φ depends on the ratio between Young's modulus and the uniaxial yield stress (E/σ_Y), as well as Poisson's ratio ν; φ is reported in literature to be 2.7–3 for most ceramic materials [44]. Assuming that $\varphi = 2.7$ and Vickers hardness equals $H_V = 0.4636 \cdot P/a^2$, Equation (1) can be written as:

$$K_{Ic} = 0.0089 \cdot \left(\frac{E}{H_V} \right)^{2/5} \cdot \frac{P}{a \cdot \sqrt{l}} \tag{2}$$

It should be noted that Equation (2) is applicable for Palmqvist cracks only. At low crack-to-indent ratios, the dominantly formed crack geometry are Palmqvist cracks, typically showing a ratio $\frac{l}{a} \lesssim 1.5$. At higher loads, crack geometry changes to halfpenny-shaped cracks (radial-median cracks) showing a ratio $\frac{l}{a} \gtrsim 2.5$. Within the range $1.5 \lesssim l/a \lesssim 2.5$, both crack systems can occur. Accordingly to the literature, Y-TZP ceramics preferentially crack in the Palmqvist rather than in the halfpenny mode [45,46].

Hardness was calculated from crack-free Vickers indentations generated at a lower applied indentation load of 4.90 N.

4. Conclusions

A two-stage fabrication approach was employed to obtain a sandwich-shaped specimen consisting of a high-alloy TRIP-steel/Al$_2$O$_3$-(3Y)-TZP composite and a (3Y)-TZP layer. The composite and ceramic specimens were separately consolidated and joined using SPS.

- No evidence of cracks or voids was observed at the composite/TZP interface.
- Limited grain growth from the initial 208 nm to 226 nm after joining at 1150 °C occurred during joining. The crystallite size and the fraction of the monoclinic phase in the Y-TZP ceramic virtually did not change.
- The hardness values and the shear force for broken SPS-bonded samples did not change during joining. Slightly increased fracture toughness of the TZP-layer with increased joining temperatures was attributed to the effect of grain size on transformation toughening.
- SPS was proven to be an effective technique for sintering and solid-state joining of ceramics and metal/ceramic composites.

Acknowledgments: The authors thank Ines Schneider and Ehud Galun for their support.

Author Contributions: A.M. designed experiments and analyzed and discussed the experimental results, S.G. designed and performed the experiments on sintering and characterization of TRIP-steel/ceramic composites, S.K. prepared samples and conducted SPS joining experiments. L.K. and N.F. supervised this work and discussed the experimental results. All authors contributed to writing and approved the final manuscript.

Conflicts of Interest: The authors declare no conflict of interest. The founding sponsors had no role in the design of the study, in the collection, analyses, or interpretation of data, in the writing of the manuscript, or in the decision to publish the results.

Abbreviations

The following abbreviations are used in this manuscript:

PSZ	partially stabilized zirconia
SPS	spark plasma sintering
TRIP	transformation-induced plasticity
TZP	tetragonal zirconia polycrystal

References

1. Hannink, R.H.J.; Kelly, P.M.; Muddle, B.C. Transformation toughening in zirconia-containing ceramics. *J. Am. Ceram. Soc.* **2004**, *83*, 461–487. [CrossRef]
2. Garvie, R.C.; Hannink, R.H.; Pascoe, R.T. Ceramic steel? *Nature* **1975**, *258*, 703–704. [CrossRef]
3. Gerberich, W.W.; Hemmings, P.L.; Zackay, V.F.; Parker, E. Interactions between crack growth and strain-induced transformation. In *Fracture*; Pratt, P.L., Ed.; Chapman and Hall: London, UK, 1969; pp. 208–305.
4. Grässel, O.; Krüger, L.; Frommeyer, G.; Meyer, L. High strength Fe–Mn–(Al, Si) TRIP/TWIP steels development–properties–application. *Int. J. Plast.* **2000**, *16*, 1391–1409. [CrossRef]
5. Prüger, S.; Kuna, M.; Wolf, S.; Krüger, L. A material model for TRIP-steels and its application to a CrMnNi cast alloy. *Steel Res. Int.* **2011**, *82*, 1070–1079. [CrossRef]
6. Kovalev, A.; Jahn, A.; Weiß, A.; Scheller, P.R. Characterization of the TRIP/TWIP effect in austenitic stainless steels using stress-temperature-transformation (STT) and deformation-temperature-transformation (DTT) diagrams. *Steel Res. Int.* **2011**, *82*, 45–50. [CrossRef]
7. Wolf, S.; Martin, S.; Krüger, L.; Martin, U. Constitutive modelling of the rate dependent flow stress of cast high-alloyed metastable austenitic TRIP/TWIP steel. *Mater. Sci. Eng. A* **2014**, *594*, 72–81. [CrossRef]
8. Martin, S.; Richter, S.; Decker, S.; Martin, U.; Krüger, L.; Rafaja, D. Reinforcing mechanism of Mg-PSZ particles in highly-alloyed TRIP steel. *Steel Res. Int.* **2011**, *82*, 1133–1140. [CrossRef]
9. Aneziris, C.G.; Schärfl, W.; Biermann, H.; Martin, U. Energy-absorbing TRIP-Steel/Mg-PSZ composite honeycomb structures based on ceramic extrusion at room temperature. *Int. J. Appl. Ceram. Technol.* **2009**, *6*, 727–735. [CrossRef]

10. Frage, N.; Kalabukhov, S.; Sverdlov, N.; Kasiyan, V.; Rothman, A.; Dariel, M.P. Effect of the spark plasma sintering (SPS) parameters and LiF doping on the mechanical properties and the transparency of polycrystalline Nd-YAG. *Ceram. Int.* **2012**, *38*, 5513–5519. [CrossRef]

11. Hayun, S.; Kalabukhov, S.; Ezersky, V.; Dariel, M.P.; Frage, N. Microstructural characterization of spark plasma sintered boron carbide ceramics. *Ceram. Int.* **2010**, *36*, 451–457. [CrossRef]

12. Hayun, S.; Paris, V.; Mitrani, R.; Kalabukhov, S.; Dariel, M.P.; Zaretsky, E.; Frage, N. Microstructure and mechanical properties of silicon carbide processed by Spark Plasma Sintering (SPS). *Ceram. Int.* **2012**, *38*, 6335–6340. [CrossRef]

13. Duan, R.-G.; Zhan, G.-D.; Kuntz, J.D.; Kear, B.H.; Mukherjee, A.K. Spark plasma sintering (SPS) consolidated ceramic composites from plasma-sprayed metastable Al_2TiO_5 powder and nano-Al_2O_3, TiO_2, and MgO powders. *Mater. Sci. Eng. A* **2004**, *373*, 180–186. [CrossRef]

14. Lanfant, B.; Leconte, Y.; Bonnefont, G.; Garnier, V.; Jorand, Y.; Le Gallet, S.; Pinault, M.; Herlin-Boime, N.; Bernard, F.; Fantozzi, G. Effects of carbon and oxygen on the spark plasma sintering additive-free densification and on the mechanical properties of nanostructured SiC ceramics. *J. Eur. Ceram. Soc.* **2015**, *35*, 3369–3379. [CrossRef]

15. Poyato, R.; Macías-Delgado, J.; García-Valenzuela, A.; González-Romero, R.; Muñoz, A.; Domínguez-Rodríguez, A. Electrical properties of reduced 3YTZP ceramics consolidated by spark plasma sintering. *Ceram. Int.* **2016**, *42*, 6713–6719. [CrossRef]

16. Niu, H.Z.; Chen, Y.F.; Zhang, D.L.; Zhang, Y.S.; Lu, J.W.; Zhang, W.; Zhang, P.X. Fabrication of a powder metallurgy Ti_2AlNb-based alloy by spark plasma sintering and associated microstructure optimization. *Mater. Des.* **2016**, *89*, 823–829. [CrossRef]

17. Rudinsky, S.; Hendrickx, P.; Bishop, D.P.; Brochu, M. Spark plasma sintering and age hardening of an Al–Zn–Mg alloy powder blend. *Mater. Sci. Eng. A* **2016**, *650*, 129–138. [CrossRef]

18. Minier, L.; le Gallet, S.; Grin, Y.; Bernard, F. A comparative study of nickel and alumina sintering using spark plasma sintering (SPS). *Mater. Chem. Phys.* **2012**, *134*, 243–253. [CrossRef]

19. Sun, Y.; Vajpai, S.K.; Ameyama, K.; Ma, C. Fabrication of multilayered Ti-Al intermetallics by spark plasma sintering. *J. Alloys Compd.* **2014**, *585*, 734–740. [CrossRef]

20. Ji, G.; Grosdidier, T.; Bernard, F.; Paris, S.; Gaffet, E.; Launois, S. Bulk FeAl nanostructured materials obtained by spray forming and spark plasma sintering. *J. Alloys Compd.* **2007**, *434–435*, 358–361. [CrossRef]

21. Ji, G.; Bernard, F.; Launois, S.; Grosdidier, T. Processing conditions, microstructure and mechanical properties of hetero-nanostructured ODS FeAl alloys produced by spark plasma sintering. *Mater. Sci. Eng. A* **2013**, *559*, 566–573. [CrossRef]

22. Seifert, M.; Shen, Z.; Krenkel, W.; Motz, G. Nb(Si,C,N) composite materials densified by spark plasma sintering. *J. Eur. Ceram. Soc.* **2015**, *35*, 3319–3327. [CrossRef]

23. Meir, S.; Kalabukhov, S.; Frage, N.; Hayun, S. Mechanical properties of Al_2O_3/Ti composites fabricated by spark plasma sintering. *Ceram. Int.* **2015**, *41*, 4637–4643. [CrossRef]

24. Liu, B.H.; Su, P.-J.; Lee, C.-H.; Huang, J.-L. Linking microstructure evolution and impedance behaviors in spark plasma sintered Si_3N_4/TiC and Si_3N_4/TiN ceramic nanocomposites. *Ceram. Int.* **2013**, *39*, 4205–4212. [CrossRef]

25. Decker, S.; Krüger, L.; Schneider, I. Influence of steel and Mg PSZ additions on the compressive deformation behavior of an Al_2O_3 reinforced TRIP/TWIP-matrix-composite. In Proceedings of the International Powder Metallurgy Congress and Exhibition, Gothenburg, Sweden, 15–18 September 2013; pp. 113–118.

26. Krüger, L.; Grützner, S.; Decker, S.; Schneider, I. Spark plasma sintering and strength behavior under compressive loading of Mg-PSZ/Al_2O_3-TRIP-steel composites. *Mater. Sci. Forum* **2015**, *825–826*, 182–188. [CrossRef]

27. Radwan, M.; Nygren, M.; Flodström, K.; Esmaelzadeh, S. Fabrication of crack-free SUS316L/Al_2O_3 functionally graded materials by spark plasma sintering. *J. Mater. Sci.* **2011**, *46*, 5807–5814. [CrossRef]

28. Decker, S.; Kruger, L. Improved mechanical properties by high-energy milling and spark plasma sintering of a TRIP-matrix composite. *J. Compos. Mater.* **2015**, *50*, 1829–1863. [CrossRef]

29. Krüger, L.; Decker, S.; Ohser-Wiedemann, R.; Ehinger, D.; Martin, S.; Martin, U.; Seifert, H.J. Strength and failure behaviour of spark plasma sintered steel-zirconia composites under compressive loading. *Steel Res. Int.* **2011**, *82*, 1017–1021. [CrossRef]

30. Decker, S.; Krüger, L.; Richter, S.; Martin, S.; Martin, U. Strain-rate-dependent flow stress and failure of an Mg-PSZ reinforced TRIP matrix composite produced by spark plasma sintering. *Steel Res. Int.* **2012**, *83*, 521–528. [CrossRef]

31. Poklad, A.; Motylenko, M.; Klemm, V.; Schreiber, G.; Martin, S.; Decker, S.; Abendroth, B.; Haverkamp, M.; Rafaja, D. Interface phenomena responsible for bonding between TRIP steel and partiallystabilised zirconia as revealed by TEM. *Adv. Eng. Mater.* **2013**, *15*, 627–637. [CrossRef]

32. Lee, G.; Yurlova, M.S.; Giuntini, D.; Grigoryev, E.G.; Khasanov, O.L.; McKittrick, J.; Olevsky, E.A. Densification of zirconium nitride by spark plasma sintering and high voltage electric discharge consolidation: A comparative analysis. *Ceram. Int.* **2015**, *41*, 14973–14987. [CrossRef]

33. Delaizir, G.; Bernard-Granger, G.; Monnier, J.; Grodzki, R.; Kim-Hak, O.; Szkutnik, P.-D.; Soulier, M.; Saunier, S.; Goeuriot, D.; Rouleau, O.; et al. A comparative study of spark plasma sintering (SPS), hot isostatic pressing (HIP) and microwaves sintering techniques on p-type Bi_2Te_3 thermoelectric properties. *Mater. Res. Bull.* **2012**, *47*, 1954–1960. [CrossRef]

34. He, D.; Fu, Z.; Wang, W.; Zhang, J.; Munir, Z.A.; Liu, P. Temperature-gradient joining of Ti–6Al–4V alloys by pulsed electric current sintering. *Mater. Sci. Eng. A* **2012**, *535*, 182–188. [CrossRef]

35. Hirose, T.; Shiba, K.; Ando, M.; Enoeda, M.; Akiba, M. Joining technologies of reduced activation ferritic/martensitic steel for blanket fabrication. *Fusion Eng. Des.* **2006**, *81*, 645–651. [CrossRef]

36. Miriyev, A.; Stern, A.; Tuval, E.; Kalabukhov, S.; Hooper, Z.; Frage, N. Titanium to steel joining by spark plasma sintering (SPS) technology. *J. Mater. Process. Technol.* **2013**, *213*, 161–166. [CrossRef]

37. Aroshas, R.; Kalabukhov, S.; Stern, A.; Frage, N. Diffusion bonding of ceramics by spark plasma sintering (SPS) apparatus. *Adv. Mater. Res.* **2015**, *1111*, 97–102. [CrossRef]

38. Liu, L.; Ye, F.; Zhou, Y.; Zhang, Z.; Hou, Q. Fast bonding α-SiAlON ceramics by spark plasma sintering. *J. Eur. Ceram. Soc.* **2010**, *30*, 2683–2689. [CrossRef]

39. Ruiz, L.; Readey, M.J. Effect of heat treatment on grain size, phase assemblage and mechanical properties of 3 mol % Y-TZP. *J. Am. Ceram. Soc.* **1996**, *79*, 2331–2340. [CrossRef]

40. Lange, F.F. Transformation toughening. *J. Mater. Sci.* **1982**, *17*, 225–234. [CrossRef]

41. Bravo-Leon, A.; Morikawa, Y.; Kawahara, M.; Mayo, M.J. Fracture toughness of nanocrystalline tetragonal zirconia with low yttria content. *Acta Mater.* **2002**, *50*, 4555–4562. [CrossRef]

42. Casellas, D.; Feder, A.; Llanes, L.; Anglada, M. Fracture toughness and mechanical strength of Y-TZP/PSZ ceramics. *Scr. Mater.* **2001**, *45*, 213–220. [CrossRef]

43. Niihara, K.; Morena, R.; Hasselman, D.P.H. Evaluation of K_{Ic} of brittle solids by the indentation method with low crack-to-indent ratios. *J. Mater. Sci. Lett.* **1982**, *1*, 13–16. [CrossRef]

44. Ponton, C.B.; Rawlings, R.D. Vickers indentation fracture toughness test Part 1 Review of literature and formulation of standardised indentation toughness equations. *Mater. Sci. Technol.* **2013**, *5*, 865–872. [CrossRef]

45. Kaliszewski, M.S.; Behrens, G.; Heuer, A.H.; Shaw, M.C.; Marshall, D.B.; Dransmanri, G.W.; Steinbrech, R.W.; Pajares, A.; Guiberteau, F.; Cumbrera, F.L.; et al. Indentation Studies on Y_2O_2-Stabilized ZrO_2: I, Development of Indentation-Induced Cracks. *J. Am. Ceram. Soc.* **1994**, *77*, 1185–1193. [CrossRef]

46. Cottom, B.A.; Mayo, M.J. Fracture toughness of nanocrystalline ZrO_2-3 mol % Y_2O_3 determined by vickers indentation. *Scr. Mater.* **1996**, *34*, 809–814. [CrossRef]

materials

MDPI

Article

Microstructure and Electrical Properties of AZO/Graphene Nanosheets Fabricated by Spark Plasma Sintering

Shuang Yang [1], Fei Chen [1,*], Qiang Shen [1], Enrique J. Lavernia [2] and Lianmeng Zhang [1]

[1] State Key Laboratory of Advanced Technology for Materials Synthesis and Processing, Wuhan University of Technology, Wuhan 430070, China; shuang_yang@yeah.net (S.Y.); sqqf@263.net (Q.S.); lmzhang195501@126.com (L.Z.)

[2] Department of Chemical Engineering and Materials Science, University of California, Davis, CA 95616, USA; lavernia@ucdavis.edu

* Correspondence: chenfei027@gmail.com; Tel.: +86-27-87217492

Academic Editor: Eugene A. Olevsky

Received: 12 May 2016; Accepted: 20 July 2016; Published: 29 July 2016

Abstract: In this study we report on the sintering behavior, microstructure and electrical properties of Al-doped ZnO ceramics containing 0–0.2 wt. % graphene sheets (AZO-GNSs) and processed using spark plasma sintering (SPS). Our results show that the addition of <0.25 wt. % GNSs enhances both the relative density and the electrical resistivity of AZO ceramics. In terms of the microstructure, the GNSs are distributed at grain boundaries. In addition, the GNSs are also present between ZnO and secondary phases (e.g., $ZnAl_2O_4$) and likely contribute to the measured enhancement of Hall mobility (up to 105.1 $cm^2 \cdot V^{-1} \cdot s^{-1}$) in these AZO ceramics. The minimum resistivity of the AZO-GNS composite ceramics is 3.1×10^{-4} $\Omega \cdot cm$ which compares favorably to the value of AZO ceramics which typically have a resistivity of 1.7×10^{-3} $\Omega \cdot cm$.

Keywords: graphene nanosheets; Al-doped-ZnO; electrical properties; spark plasma sintering

---

1. Introduction

The Li-ion batteries using liquid or organic electrolytes have been widely investigated for the application in electric vehicles(EV) and backup uninterruptible power supply (UPS) systems. Challenges in terms of scale-up and guaranteed safety have been proposed by many researchers. Therefore, all-solid-state batteries (ASSB) have been designed which show significant improvements in energy density and safety relative to those of commonly used organic liquid and polymer-based lithium-ion batteries. In a related work, Baek et al. [1] fabricated an all-solid-state lithium rechargeable battery with both the electrode and electrolyte processed by spark plasma sintering (SPS). The battery is characterized by a carefully designed architecture with a good lithium diffusion percolation path and low solid contact resistance without any defects and undesirable reactions arising from the sintering method. The more widely used solid-state synthesis of ABBS attaches much importance to synthesizing cathode materials.

As a result of the excellent electrical and optical properties [2], transition metal oxide–based materials have found widespread applications in electrode active materials, such as Ni-MH cells, lithium cells, solar cells, and fuel cells, due to their high electrochemical activity [3]. Recently, transition metal oxides have evolved as potential electrode materials for lithium-based batteries due to their higher theoretical capacity and non-reactive properties relative to those of conventional carbon-based materials. Among the transition metal oxides, ZnO possesses numerous advantages, such as low cost, manufacturability and environmental inertness [4]. As such, zinc oxide is a promising substitute for conventional graphite anodes in lithium-ion batteries due to its superior theoretical capacity (978 mAh·g^{-1}) which compares favorably to the theoretical capacity of graphite (372 mAh·g^{-1}) [5].

ZnO is a well-known wide band gap (3.37 eV) semiconductor with a large binding energy (60 meV) and a proven capacity for reversible electrochemical Li storage [6,7]. Feng et al. [8] reported that ZnO nanoplates used as anodes for Li-ion batteries exhibited a very high capacity of ~680 mAh·g^{-1}, with continuous fading for a stable capacity of ~180 mAh·g^{-1}. Recently, Bresser et al. [9] utilized ZnO nanoparticles as anode materials for lithium-ion batteries and reported a reversible capacity of ~700 mAh·g^{-1} in the first cycle, but it rapidly faded during subsequent charge-discharge cycles before stabilizing at ~260 mAh·g^{-1}. Hence, on the basis of these published results, it is evident that ZnO is an attractive candidate material for anodes because of its inherent properties, which include low reversible capacity and a severe capacity fading response. There are, however, important technical barriers that need to be addressed before ZnO can be successfully used as anodes, such as a low electrical conductivity, which hinders the practical application of ZnO-based anodes in lithium-ion batteries [10].

When used as anodes for Li-ion batteries, graphene-based metal oxides display significantly improved performance in terms of cycling stability and rate capability [11–14]. Furthermore, published results suggest that graphene nanosheets may enhance the electronic conductivity of the overall electrode, as well as stabilize the microstructure of metal oxides during cycles [15–17]. These effective methods have also shown much impact on the improvement of the electrochemical performance for ZnO. In one study, metal-doped ZnO nanoparticles and carbon-based composites (e.g., carbon-coated ZnO nanorods) [18,19] were prepared for use as anodes in lithium-ion batteries. Moreover, other studies have shown that the electrochemical performance is significantly improved by the substitution of zinc by iron within the wurtzite lattice with a reversible capacity of ~410 mAh·g^{-1} after 100 cycles compared to pure zinc oxide of ~260 mAh·g^{-1} [9]. In addition, in a related study, a Li-ion cell assembly was prepared consisting of a ZnO multiwalled carbon nanotube nanocomposite free-standing anode and a Li metal cathode, and the results reveal an excellent discharge capacity, remaining as high as 460 mAh·g^{-1} after 100 cycles [20]. Thus, it is evident that ZnO-based ceramics should be investigated as promising anode materials for all-solid-state lithium rechargeable batteries.

In view of the above discussion, the present work was motivated by the following three questions. First, what is the feasibility of using spark plasma sintering (SPS) to fabricate AZO-GNSs composite ceramics? Second, what are the corresponding densification behavior, microstructure characteristics and electrical properties? Third, what are the underlying mechanisms that govern the electrical response and what is the influence, if any, of the spatial distribution of GNS?

2. Experimental Procedures

Graphene nanosheets produced using a modified Hummer's method, were purchased from NANO XF (Nanjing, China). The resultant platelets were approximately 0.8–1.2 nm in thickness and 0.5–2 mm in length. As-received AZO nanoparticles (Huzheng Nanotechnology Co. Ltd., Shanghai, China) were used in this study. The composition of the AZO nanoparticles is Al_2O_3/ZnO of 2/98 (mol %); purity > 99.9%; specific surface area of 30–50 m^2/g; apparent density of 1.21 g/cm^3; average particle size of 80~100 nm; and resistivity of 1.2×10^{-1} Ω·m.

The as-received GNSs were dispersed in a solution of dimethyl formamide (DMF) with ultrasonic agitation for 1 h. Then, AZO powders were added to these slurries at GNSs ratios of 0, 0.025, 0.05, 0.1 and 0.2 wt. %, and further sonicated for 30 min. The compounds were then milled at 300 rpm by a planetary ball mill (QM-3SP2, Nanjing, China) for 24 h with a weight ratio of ball to powder of 3:2, and DMF as a solvent. The composite powders were then dried at 100 °C using a vacuum drying oven for 24 h. The composite powders were loaded into a 20-mm-diameter graphite die inside the SPS furnace (SPS-1050, Tokyo, Japan). A piece of graphitic paper was placed between the punch and powder in order to facilitate removal of the sintered sample. Subsequently, the composite powders were sintered at 900 °C, 1000 °C, 1100 °C and 1200 °C from room temperature with a temperature rate of 100 °C/min. The applied uniaxial pressure was 40 MPa and dwell time was 3 min under a vacuum of 20 Pa. The temperature was monitored using an infrared instrument during the sintering process. The obtained bulk with the GNSs content of 0, 0.025, 0.05, 0.1 and 0.2 are hereby designated as AZO, AZO-0.025G, AZO-0.05G, AZO-0.1G and AZO-0.2G, respectively.

The density of AZO-GNSs composite ceramics were measured via Archimedes' method using deionized water immersion, and the relative density was calculated by taking the theoretical density of AZO as 5.67 g/cm^3. The carrier concentration, Hall mobility and resistivity of AZO ceramics were measured at room temperature using the Hall measurement system (Accent, HL5500PC, Hertfordshire, UK) according to the van der Pauw method [21]. The microstructures of the sintered ceramics were observed by field emission scanning electron microscopy (FESEM, 20 kV, FEI Quanta FEG250, Hillsboro, OR, USA) and a high-resolution transmission electron microscopy (JEM-2100F STEM, Tokyo, Japan), The average grain size was measured by the intercept-line method, using 1.225 as the stereological correction factor. Approximately 200 grains were used for each data set reported [22].

3. Results

3.1. Phase Composition and Microstructure of AZO-G Composite Nanoparticles

Figure 1a shows the XRD patterns for AZO-0.05G composites nanoparticles following ball milling for 24 h. It can be observed that all diffraction peaks are consistent with those of ZnO (PDF #36-1451), with a quartzite structure indicating only one crystalline phase related to ZnO. The peaks of GNSs in XRD patterns are not observed due to the ultra-low percentage of GNSs in the as-prepared composite powders. A macroscopic view of the AZO nanoparticles after ball milling with 0.025 wt. % GNSs is evident in Figure 2a. The grain size of AZO-0.025G composite nanoparticles is approximately 100 nm. The thin transparent GNSs are not observed in AZO-0.025G composite powders and there are no obvious agglomerations in the composite powder after ultrasonication and ball milling. The microstructure of AZO-0.05G and AZO-0.1G composite powders is shown in Figure 2b,c. In the case of the AZO-0.05G composite powders, the transparency indicates that the thin platelets contained only a few graphene nanosheets. The graphene nanosheets are well distributed throughout the AZO-0.05G powder and reveal no agglomeration. However, in the case of a high volume of GNSs, as in AZO-0.1G, agglomerates of graphene nanosheets were observed in some regions. An example of large agglomerates of segregated GNSs in AZO-0.1G composite powder is shown in Figure 2c by the arrow.

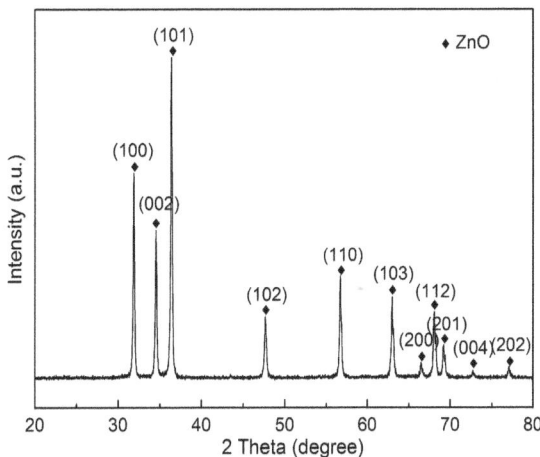

Figure 1. XRD patterns of AZO-0.05G composite powder.

Figure 2. SEM images of composite powders: (**a**) AZO-0.025G; (**b**) AZO-0.05G; (**c**) AZO-0.1G.

3.2. Relative Density and Grain Size of Composite Ceramics

Table 1 shows the relative density and grain size of the AZO-G composite ceramics for different sintering temperatures and with different GNSs contents. The relative density of pure AZO ceramic increases as the sintering temperature varies from 1000 to 1200 °C and it attains a maximum value at the sintering temperature of 1200 °C. However, in the case of the AZO-0.025G composite ceramics, the relative density increases from 98.3% to 99.2% when the sintering temperature increases from 1000 to 1100 °C. Increasing the temperature to 1200 °C, however, leads to a decrease in the relative density of the AZO-0.025G composite ceramics. Furthermore, the AZO-G composite ceramics with 0.025 wt. % GNSs attain the highest relative density at the sintering temperature of 1100 °C. It can be seen from Table 1 that the relative density of AZO-G composite ceramics decreases slightly with the increase of GNSs, and this is attributed to the agglomeration of the graphene nanosheets.

The AZO-0.025G composite ceramic exhibits limited grain refinement, as shown in Table 1. The average grain size of AZO-0.025G composite ceramics sintered at 1100 °C is 2.8 μm. A well-known response of GNSs during sintering is a phenomenon described as grain wrapping [23], where GNSs act as diffusion barriers and hence effectively pin the grains by wrapping themselves around them. In the case of lower volume fractions of GNSs (e.g., AZO-0.025G), only a few grains are wrapped and pinned. Although partial grain pinning may lead to a decrease in the grain size, it also promotes abnormal grain growth in AZO. This phenomenon, which leads to grain refinement in some regions and abnormal grain growth in others, has been reported in other studies [24]. From Reference [24], it is clear that Al_2O_3/graphene composites show no significant grain refinement with the graphene volume fraction of 5%. However, with the increase of the volume fraction of graphene from 5% to 15%, the grain size for Al_2O_3/graphene has been significantly decreased. As another example, TaC/graphene composites [25] exhibit a limited grain refinement with the addition of graphene (1 vol. %) and a sharply decreased grain size of TaC/graphene composites with the addition of graphene (3 and 5 vol. %). Similarly, the extent of grain refinement is significant when the volume fraction of GNSs is increased in our work. The average grain size of AZO-0.05G and AZO-0.1G composite ceramics is 1.5 and 0.9 μm, respectively. In this case, a high volume fraction of GNSs will effectively stabilize the grains as a higher percentage of them are "wrapped".

Table 1. Relative density and average grain size of the sintered samples.

GNSs (wt. %)	Temperature (°C)	Relative Density (%)	Average Grain Size (μm)
0	1000	97.8	2.1
0	1100	98.4	3.3
0	1200	99.3	5.3
0.025	1000	98.3	1.8
0.025	1100	99.2	2.8
0.025	1200	97.2	4.3
0.05	1100	98.3	1.5
0.1	1100	97.2	0.9

3.3. Microstructure of AZO-G Composite Ceramics

The SEM micrographs of the fracture surface of the AZO-0.05G composite ceramics sintered at various temperatures are shown in Figure 3. As seen in the figure, the distribution of GNSs is homogenous within the AZO ceramic matrix for the range of sintering temperatures studied, 1000 °C, 1100 °C and 1200 °C. It is clear that the microstructure of AZO-0.05G is dense following sintering at 1000 °C. There is, however, some distributed porosity that can be seen on the fracture surfaces corresponding to a sintering temperature of 1200 °C, indicating a low relative density for the AZO-0.05G composite ceramics.

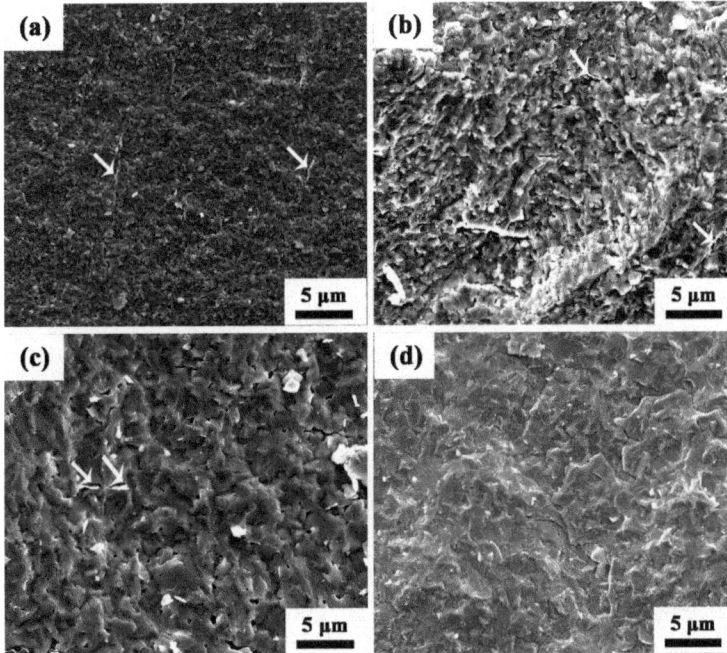

Figure 3. SEM micrographs of AZO-0.05G composite ceramics at sintering temperatures of: (a) 1000 °C; (b) 1100 °C; (c) 1200 °C; (d) 1300 °C.

Figure 4 shows the fracture surface corresponding to the AZO ceramic and AZO-G composite ceramics with various GNS contents, respectively. The results show that uniformly distributed GNSs in the AZO matrix (highlighted by white arrows) have no apparent influence on the relative density of AZO-G composite ceramics as the GNS content varies from 0.025 to 0.05 wt. %. For AZO-0.1G and AZO-0.2G, however, significant cracks around graphene nanosheets are observed in Figure 4f due to the agglomeration of GNSs, ultimately affecting the relative density of the AZO-G composite ceramics shown in Table 1.

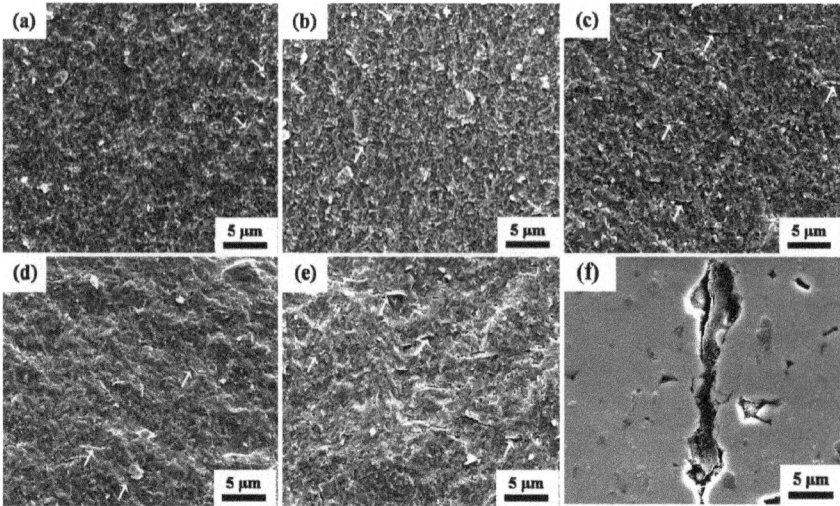

Figure 4. SEM micrographs of AZO-G ceramics with GNS content of (**a**) 0 wt. %; (**b**) 0.025 wt. %; (**c**) 0.05 wt. %; (**d**) 0.1wt. %; (**e**) 0.2 wt. % and (**f**) 0.2 wt. % (polished surface).

Figure 5 shows HRTEM images of the spatial distribution of GNSs in the AZO matrix. As seen from Figure 5a, GNSs with a thickness of about 10 nm are located between ZnO/ZnO grain boundaries and the interface between ZnO and graphene appears to be free from other impurities. Another unique and significant characteristic in the microstructure of AZO-G composite ceramics shown in Figure 5b is that, although GNSs are generally flat and straight at ZnO/ZnO grain boundaries, they are also observed at a ZnO/ZnAl$_2$O$_4$ grain boundary and are connected to each other to form a network at a grain boundary triple junction.

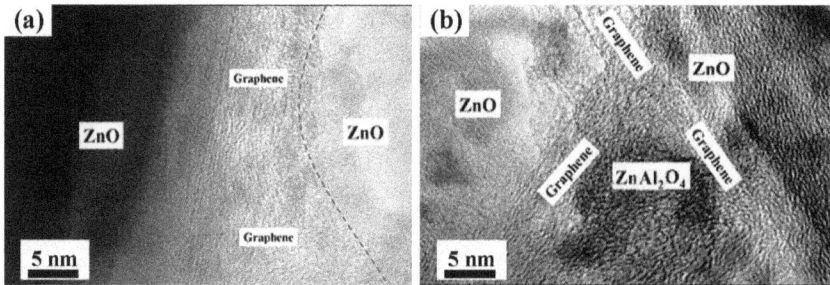

Figure 5. HRTEM image of AZO-0.05G composite ceramics: (**a**) GNSs at ZnO/ZnO grain boundary; (**b**) GNSs at ZnO/ZnAl$_2$O$_4$ grain boundary and grain boundary triple junction.

3.4. Electrical Properties of AZO-G Composite Ceramics

Figure 6a shows the room temperature resistivity, carrier concentration and Hall mobility for the AZO-0.05G composite ceramics as a function of the sintering temperature. The results show a decrease of resistivity for AZO-0.05G composite ceramics as the sintering temperature is increased from 1000 to 1200 °C. However, the resistivity of AZO-0.05G composite ceramics exhibits a sharp increase at the sintering temperature of 1300 °C. The deterioration of resistivity is attributed to the oxidation of GNSs in AZO-0.05G, which is apparent in Figure 4d. It is clear that the resistivity of AZO-0.05G composite ceramics sintered at 1100 °C attains a minimum value of 4.1×10^{-4} $\Omega \cdot$ cm, as shown in Figure 6b.

As the content of GNSs ranges from 0.05 to 0.2 wt. %, the resistivity and Hall mobility reveal a increase and sharp decrease, respectively, which can be explained on the basis of a deterioration in relative density caused by the agglomeration of graphene nanosheets.

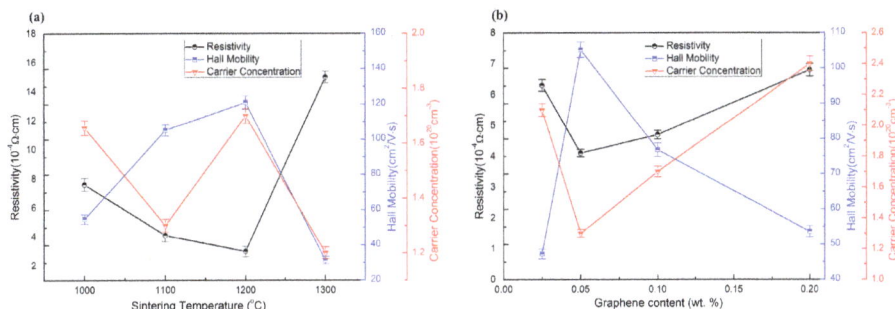

Figure 6. The electrical property of AZO-G composite ceramics: (**a**) at various sintering temperature; (**b**) with various graphene nanosheets content.

4. Discussion

4.1. Influence of GNSs on Densification of AZO-GNSs during SPS

Table 1 shows that the addition of the appropriate amount of GNSs increases the relative density of AZO ceramics. However, the relative density of the AZO-0.025G composite ceramics exhibits a sharp increase as the sintering temperature increases to 1300 °C. To provide insight into the influence of GNSs on the sintering behavior of AZO ceramics, shrinkage as a function of the SPS sintering temperature for both the AZO ceramics and the AZO-0.025G composite ceramics is examined in the present study. Figure 7 shows the normalized densification behavior of the AZO and AZO-0.025G ceramics sintered at 100 °C/min and 40 MPa. The density values were calculated from the shrinkage curves by correcting for the thermal expansion of the graphite die, which was separately measured using dummy tests at the respective SPS conditions. As illustrated, the spark plasma sintering of AZO ceramics begins to shrink at ~400 °C and is complete at ~1200 °C, and hence follows a one-step densification. In contrast, the sintering behavior of AZO-0.025G shows a two-step shrinkage response. The first shrinkage occurs between ~600 and ~1100 °C (denoted as region A) and the second stage starts at ~1250 °C (denoted as region B). It is worth mentioning that the density of AZO-0.025G composite ceramics decreases in the sintering range from 1100 to 1200 °C, which is consistent with the porosity observed in AZO-0.025G at a sintering temperature of 1200 °C shown in Figure 3c. The underlying mechanism that results in the loss of GNSs at a higher sintering temperature is explained as follows. Graphene nanosheets can react with O_2 to form CO or CO_2, because Al doping is supposed to lead to an increase of the charge carrier concentration following Equation (1) [26]:

$$Al_2O_3 \xrightarrow{ZnO} 2Al^{\bullet}_{Zn} + 2e' + \frac{1}{2}O_2 \uparrow \qquad (1)$$

When increasing the sintering temperature from 1200 to 1300 °C, graphene nanosheets are completely oxidized by oxygen from Equation (1). Thus, the porosity in AZO-0.025G composite ceramics decreases as the sintering temperature increases to 1300 °C. Although shrinkage of AZO ceramics occurs at a lower temperature of 400 °C, it is observed that fully dense AZO-0.025G composite ceramics are obtained at a lower temperature at 1100 °C, which is 100 °C lower than that required by AZO ceramics. With the addition of GNSs, significant increases in densification have been also observed for conductive ceramics such as TaC [25] and ZrB_2 [27]. The higher conductivity enables significant current to flow directly through the sample during the SPS process and hence these ceramics benefit from both the increased thermal and electrical conductivity of graphene.

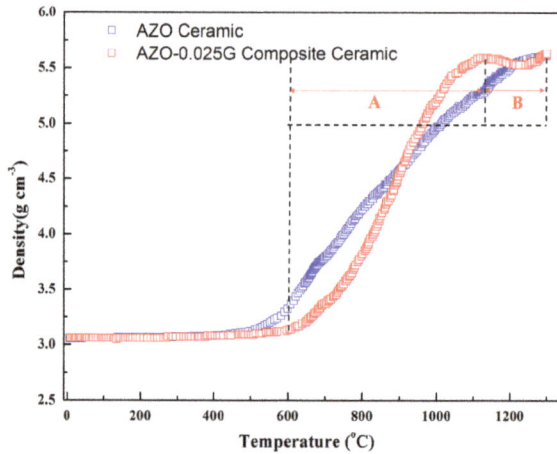

Figure 7. Normalized shrinkage of AZO ceramic and AZO-0.025G composite ceramics.

4.2. Influence of GNSs on Electrical Properties of AZO-G

The resistivity of AZO ceramics is primarily determined by the carrier concentration (*n*) and Hall mobility (*μ*), as shown by the following equation:

$$\rho \equiv \frac{1}{n\mu e} \qquad (2)$$

where *e* is a constant of electron charge ($e = 1.602 \times 10^{-19}$ C), n is the carrier concentration, *μ* is the Hall mobility. It is obvious that the resistivity of AZO ceramics is determined by the carrier concentration and Hall mobility, which can be seen from Equation (2). The carrier concentration (*n*) of AZO is determined by the substitution of Al^{3+} on Zn^{2+} and oxygen loss [28]. Hall mobility is influenced by impurity scattering (μ_i), grain boundary scattering (μ_g) and lattice thermal vibration scattering (μ_l), as shown in the following expression [29,30]:

$$\frac{1}{\mu} = \frac{1}{\mu_i} + \frac{1}{\mu_g} + \frac{1}{\mu_l} \qquad (3)$$

The thermal vibration scattering effect is very weak at room temperature and can be neglected. Therefore, impurity scattering and grain boundary scattering are the main factors that limit Hall mobility. As the sintering temperature exceeds 1000 °C, Al cannot fully dissolve into the ZnO structure as the solubility of Al in ZnO is very small [28]. Thus, further Al additions leads to the formation of the non-conductive $ZnAl_2O_4$ phase and the $ZnAl_2O_4$ is segregated to grain boundaries and grain boundary triple junctions, which effectively decreases the electrical and thermal conductivities [26].

Figure 8 shows the relationship among resistivity, Hall mobility and carrier concentration for AZO and AZO-0.05G composite ceramics at various sintering temperatures. The AZO-0.05G composite ceramics show a fairly low resistivity of 3.1×10^{-4} Ω·cm, while the lowest resistivity of 1.7×10^{-3} Ω·cm for AZO ceramics is one order of magnitude than that of AZO-0.05G composite ceramics. Moreover, the addition of GNSs for AZO-0.05G composite ceramics significantly enhances the Hall mobility with a value from 27.4 to 105.1 $cm^2 \cdot V^{-1} \cdot s^{-1}$ which effectively enhances the resistivity.

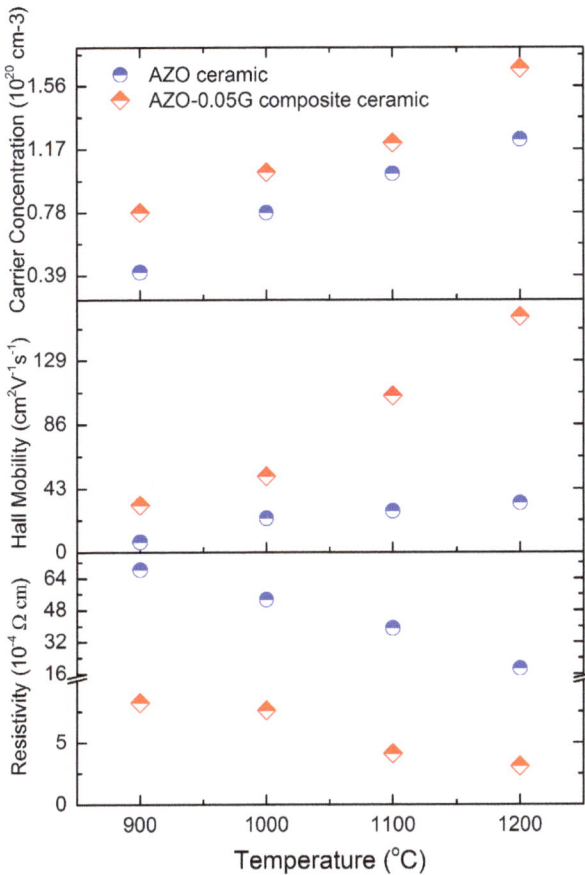

Figure 8. The resistivity, Hall mobility and carrier concentration for AZO ceramics and AZO-0.05G composite ceramics.

In the case of AZO-G composite ceramics, the enhancement of electrical conductivity is principally attributable to the addition of GNSs in the AZO matrix. To that effect, there are two factors related to the enhancement of the conductivity of AZO-G composite ceramics: (1) the significantly higher electron mobility of GNSs relative to that of AZO; and (2) an efficient and integrated conductive network of GNSs in the copper matrix. In other words, when the volume fraction of GNSs is low, it is not possible to attain the continuous network needed to result in an overall increase in electrical conductivity. However, if the GNS content is too high, agglomeration can lead to microstructural flaws such as the formation of pores and cracks, which contribute to carrier scattering, thereby leading to an electrical conductivity decrease. Moreover, it is worth noting that the formation of the spinel-phase $ZnAl_2O_4$ contributes to a high degree of impurity scattering, both at grains and grain boundaries [31], which can be both significant and difficult to ameliorate [28]. Our work shows that by adding GNSs to an AZO matrix, ZnO/ZnO and $ZnO/ZnAl_2O_4$ grain boundaries play a role in the overall electrical response. For example, the grain boundary triple junction architecture shown in Figure 5 suggests that GNSs can provide efficient electron transport pathways in local regions including those associated with non-conductive $ZnAl_2O_4$ grains.

5. Conclusions

AZO-G composite ceramics with uniformly dispersed GNSs have been prepared via ball milling using dimethyl formamide followed by SPS. In the case of AZO-0.025G, a high relative density of 99.2% was attained at a sintering temperature of 1100 °C, which is comparable to AZO ceramics sintered at a sintering temperature of 1200 °C. The resistivity of AZO-0.05G composite ceramics sintered at 1200 °C shows a minimum value of 3.1×10^{-4} $\Omega \cdot$ cm with a superior Hall mobility of 105.1 $cm^2 \cdot V^{-1} \cdot s^{-1}$. For pure AZO ceramics, the resistivity was 1.7×10^{-3} $\Omega \cdot$ cm and the Hall mobility was 27.4 $cm^2 \cdot V^{-1} \cdot s^{-1}$. The improved electrical resistivity of the AZO-G composite ceramics is attributable to the distribution of GNSs at ZnO/ZnO grain boundaries, $ZnO/ZnAl_2O_4$ grain boundaries, and the grain boundary triple junctions. Moreover, the network of GNSs provides efficient electron transport pathways in local regions, including those associated with non-conductive $ZnAl_2O_4$ grains.

Acknowledgments: This work is supported by the International Science & Technology Cooperation Program of China (No. 2011DFA52650), the National Natural Science Foundation of China (No. 51202171, No. 51521001), the Fundamental Research Funds for the Central Universities in China, and the 111 Project (B13035).

Author Contributions: Shuang Yang and Fei Chen conceived and designed the experiments; performed density and microstructure investigations; analyzed the data; and wrote the paper; Enrique J. Lavernia supervised and revised the article; Qiang Shen and Lianmeng Zhang participated in the discussion.

Conflicts of Interest: The authors declare no conflict of interest.

References

1. Seung, W.B.; Jae, M.L.; Tae, Y.K.; Min, S.S.; Youngsin, P. Garnet related lithium ion conductor processed by spark plasma sintering for all solid state batteries. *J. Power Sources* **2014**, *249*, 197–206.

2. Lee, J.H.; Ko, K.H.; Park, B.O. Electrical and optical properties of ZnO transparent conducting films by the sol-gel method. *J. Cryst. Growth* **2003**, *247*, 119–125. [CrossRef]

3. Yuan, Y.F.; Tu, J.P.; Wu, H.M.; Yang, Y.Z.; Shi, D.Q.; Zhao, X.B. Electrochemical performance and morphology evolution of nanosized ZnO as anode material of Ni-Zn batteries. *Electrochim. Acta* **2006**, *51*, 3632–3636. [CrossRef]

4. Xie, Q.; Zhang, X.; Wu, X.; Wu, H.; Liu, X.; Yue, G.; Yang, Y.; Peng, D.L. Yolk-shell ZnO-C microspheres with enhanced electrochemical performance as anode material for lithium ion batteries. *Electrochim. Acta* **2014**, *125*, 659–665. [CrossRef]

5. Guler, M.O.; Cetinkaya, T.; Tocoglu, U.; Akbulut, H. Electrochemical performance of MWCNT reinforced ZnO anodes for Li-ion batteries. *Microelectron. Eng.* **2014**, *118*, 54–60. [CrossRef]

6. Subramanian, V.; Karki, A.; Gnanasekar, K.I.; Fannie, P.E.; Rambabu, B. Nanocrystalline TiO_2 for Li-ion batteries. *J. Power Sources* **2006**, *159*, 186–192. [CrossRef]

7. Kavan, L.; Kalbác, M.; Zukalová, M.; Exnar, I.; Lorenzen, V.; Nesper, R.; Graetzel, M. Lithium storage in nanostructured TiO_2 made by hydrothermal growth. *Chem. Mater.* **2004**, *16*, 477–485. [CrossRef]

8. Li, F.; Yang, L.; Xu, G. Hydrothermal self-assembly of hierarchical flower-like ZnO nanospheres with nanosheets and their application in Li-ion batteries. *J. Alloy. Compd.* **2013**, *577*, 663–668. [CrossRef]

9. Bresser, D.; Mueller, F.; Fiedler, M. Transition-metal-doped zinc oxide nanoparticles as a new lithium-ion anode material. *Chem. Mater.* **2013**, *25*, 4977–4985. [CrossRef]

10. Chae, O.B.; Park, S.; Ryu, J.H.; Oh, S.M. Flower-like ZnO–NiO–C films with high reversible capacity and rate capability for lithium-ion batteries. *Electrochem. Soc.* **2013**, *160*, 11–14. [CrossRef]

11. Park, S.K.; Jin, A.; Yu, S.H.; Ha, J.; Jang, B.; Bong, S.; Woo, S.; Sung, Y.E.; Piao, Y. *In Situ* Hydrothermal Synthesis of Mn_3O_4 Nanoparticles on Nitrogen-doped Graphene as High-Performance Anode materials for Lithium Ion Batteries. *Electrochim. Acta* **2014**, *120*, 452–459. [CrossRef]

12. Zhou, X.Y.; Zhang, J.; Su, Q.M.; Shi, J.J.; Liu, Y.; Du, G.H. Nanoleaf-on-sheet CuO/graphene composites: Microwave-assisted assemble and excellent electrochemical performances for lithium ion batteries. *Electrochim. Acta* **2014**, *125*, 615–621. [CrossRef]

13. Yue, W.B.; Yang, S.; Ren, Y.; Yang, X.J. In situ growth of Sn, SnO on graphene nanosheets and their application as anode materials for lithium-ion batteries. *Electrochim. Acta* **2013**, *92*, 412–420. [CrossRef]

14. Yue, W.B.; Jiang, S.H.; Huang, W.J.; Gao, Z.Q.; Li, J.; Ren, Y.; Zhao, X.H.; Yang, X.J.J. Sandwich-structural graphene-based metal oxides as anode materials for lithium-ion batteries. *Mater. Chem. A* **2013**, *1*, 6928–6933. [CrossRef]

15. Yue, W.B.; Lin, Z.Z.; Jiang, S.H.; Yang, X. Preparation of graphene-encapsulated mesoporous metal oxides and their application as anode materials for lithium-ion batteries. *Mater. Chem.* **2012**, *22*, 16318–16323. [CrossRef]

16. Zhao, L.; Yue, W.B.; Ren, Y. Synthesis of graphene-encapsulated mesoporous In_2O_3 with different particle size for high-performance lithium storage. *Electrochim. Acta* **2014**, *116*, 31–38. [CrossRef]

17. Liu, J.P.; Li, Y.Y.; Ding, R.M.; Jiang, J.; Hu, Y.Y.; Ji, X.X.; Chi, Q.B.; Zhu, Z.H.; Huang, X.T. Carbon/ZnO nanorod array electrode with significantly improved lithium storage capability. *J. Phys. Chem. C* **2009**, *113*, 5336–5339. [CrossRef]

18. Shen, X.Y.; Mu, D.B.; Chen, S.; Wu, B.R.; Wu, F. Enhanced Electrochemical Performance of ZnO-Loaded/Porous Carbon Composite as Anode Materials for Lithium Ion Batteries. *ACS Appl. Mater. Interfaces* **2013**, *5*, 3118–3125. [CrossRef] [PubMed]

19. Köse, H.; Karaal, Ş.; Aydın, A.O. A facile synthesis of zinc oxide/multiwalled carbon nanotube nanocomposite lithium ion battery anodes by sol–gel method. *J. Power Sources* **2015**, *295*, 235–245. [CrossRef]

20. Yang, S.; Yue, W.B.; Zhu, J.; Ren, Y.; Yang, X.J. Graphene-based mesoporous SnO_2 with enhanced electrochemical performance for Lithium-Ion batteries. *Adv. Funct. Mater.* **2013**, *23*, 3570–3576. [CrossRef]

21. Neves, N.; Lagoa, A.; Calado, J. Al-doped ZnO nanostructured powders by emulsion detonation synthesis—Improving materials for high quality sputtering targets manufacturing. *J. Eur. Ceram. Soc.* **2014**, *34*, 2325–2338. [CrossRef]

22. Mendelson, M.I. Average grain size in polycrystalline ceramics. *J. Am. Ceram. Soc.* **1969**, *52*, 443–446. [CrossRef]

23. Walker, L.S.; Marotto, V.R.; Rafiee, M.A.; Koratkar, N.; Corral, E.L. Toughening in graphene ceramic composites. *ACS Nano* **2011**, *5*, 3182–3190. [CrossRef] [PubMed]

24. Nieto, A.; Huang, L.; Han, Y.H. Sintering behavior of spark plasma sintered alumina with graphene nanoplatelet reinforcement. *Ceram. Int.* **2015**, *41*, 5926–5936. [CrossRef]

25. Andy, N.; Debrupa, L.; Arvind, A. Graphene NanoPlatelets reinforced tantalum carbide consolidated by spark plasma sintering. *Mater. Sci. Eng. A* **2013**, *582*, 338–346.

26. Bérardan, D.; Byl, C.; Dragoe, N. Influence of the preparation conditions on the thermoelectric properties of Al-doped ZnO. *J. Am. Ceram. Soc.* **2010**, *93*, 2352–2358. [CrossRef]

27. Govindaraajan, B.; Sriharsha, K.; Arif, R.; Raman, P.S.; Kalkan, A.K.; Harimkar, S.P. Spark plasma sintering of graphene reinforced zirconium diboride ultra-high temperature ceramic composites. *Ceram. Int.* **2013**, *39*, 6637–6646.

28. Ma, N.; Li, J.F.; Zhang, B.P. Microstructure and thermoelectric properties of $Zn_{1-x}Al_xO$ ceramics fabricated by spark plasma sintering. *J. Phys. Chem. Solids* **2010**, *71*, 1344–1349. [CrossRef]

29. Zhang, D.H.; Ma, H.L. Scattering mechanisms of charge carriers in transparent conducting oxide films. *Appl. Phys. A* **1996**, *62*, 487. [CrossRef]

30. Ellmer, K.; Mientus, R. Carrier transport in polycrystalline ITO and ZnO: Al II: the influence of grain barriers and boundaries. *Thin Solid Films* **2008**, *516*, 5829–5835. [CrossRef]

31. Yang, S.; Chen, F.; Shen, Q.; Zhang, L.M. Microstructure and electrical property of aluminum doped zinc oxide ceramics by isolating current under spark plasma sintering. *J. Eur. Ceram. Soc.* **2016**, *36*, 1953–1959. [CrossRef]

materials

MDPI

Article

Zirconium Carbide Produced by Spark Plasma Sintering and Hot Pressing: Densification Kinetics, Grain Growth, and Thermal Properties

Xialu Wei [1],*, Christina Back [2], Oleg Izhvanov [2], Christopher D. Haines [3] and Eugene A. Olevsky [1,4]

[1]   Department of Mechanical Engineering, San Diego State University, 5500 Campanile Dr.,
      San Diego, CA 92182, USA; eolevsky@mail.sdsu.edu
[2]   General Atomics, 3350 General Atomics Ct., San Diego, CA 92121, USA; tina.back@ga.com (C.B.);
      oleg.izhvanov@ga.com (O.I.)
[3]   US Army Armament Research Development Engineering Center, Picatinny Arsenal, NJ 07806, USA;
      christopher.d.haines2.civ@mail.mil
[4]   Department of NanoEngineering, University of California, San Diego, 9500 Gilman Dr.,
      La Jolla, CA 92037, USA; eolevsky@ucsd.edu
*   Correspondence: xwei@mail.sdsu.edu; Tel.: +1-619-594-4627

Academic Editor: Jai-Sung Lee
Received: 15 June 2016; Accepted: 8 July 2016; Published: 14 July 2016

Abstract: Spark plasma sintering (SPS) has been employed to consolidate a micron-sized zirconium carbide (ZrC) powder. ZrC pellets with a variety of relative densities are obtained under different processing parameters. The densification kinetics of ZrC powders subjected to conventional hot pressing and SPS are comparatively studied by applying similar heating and loading profiles. Due to the lack of electric current assistance, the conventional hot pressing appears to impose lower strain rate sensitivity and higher activation energy values than those which correspond to the SPS processing. A finite element simulation is used to analyze the temperature evolution within the volume of ZrC specimens subjected to SPS. The control mechanism for grain growth during the final SPS stage is studied via a recently modified model, in which the grain growth rate dependence on porosity is incorporated. The constant pressure specific heat and thermal conductivity of the SPS-processed ZrC are determined to be higher than those reported for the hot-pressed ZrC and the benefits of applying SPS are indicated accordingly.

Keywords: zirconium carbide; spark plasma sintering; finite element simulation; grain growth; thermal properties

1. Introduction

Spark plasma sintering (SPS), also known as field-assisted sintering or current-assisted sintering, is currently one of the most attractive rapid powder consolidation techniques. It has been evidenced that the Joule heating and the hydraulic loading acting in a SPS system allow the production of dense materials at lower temperatures and during shorter periods of time compared to SPS' conventional counterpart technique—hot pressing [1–4]. Recently, SPS has been successfully utilized to consolidate ultra-high temperature ceramic (UHTC) powders, such as tantalum carbide [5], hafnium diboride [6], vanadium carbide [7], zirconium carbide [8], etc., into bulk articles with high densities and excellent properties. In addition to enhancing densification kinetics, the benefits from carrying out SPS of refractory powder-based materials include an impurities cleaning effect [9], early neck formation due to local overheating [10,11], and electric field-assisted grain size retention [12].

The aforementioned zirconium carbide (ZrC) is a typical UHTC possessing good high-temperature mechanical properties, excellent electrical and thermal conductivity, high melting point, and strong chemical resistance. It has been recently considered to be a promising candidate for high-temperature applications, such as furnace heating elements, plasma arc electrodes and nuclear cladding materials [13–15]. Although the implementations of these applications are still in progress, the attempts to consolidate ZrC powder started in the 1970s, when free-sintering and hot-pressing were employed for this purpose [16,17]. Due to ZrC's high melting point (~3500 °C) and the inherent nature of the covalent Zr–C bonding, extremely high temperatures and long-term dwellings were usually required to obtain dense ZrC products via these techniques [18,19]. In spite of the inefficiencies, these conventional consolidation approaches have been often utilized in recent years [20,21].

Investigations on SPS of ZrC were initiated with retrieving high-density specimens under moderate conditions which had never been adopted previously in free-sintering or hot-pressing of ZrC. Sciti et al. reported that up to 98% relative density could be achieved at 2100 °C under 65 MPa within 3 min when conducting SPS of micron-grade ZrC powders [22]. Submicrometric zirconium oxy-carbide (ZrC_xO_y) powders were synthesized and consolidated by Gendre et al. at about 2000 °C [23], while the vacancies introduced by the carboreduction synthesis of such powders were considered to be the factors to facilitate densification [24]. Further enhancements of densification were implemented by employing post-processed nano ZrC powders in the SPS, in which the maximum processing temperatures could be way lower than 2000 °C [25,26]. These studies have suggested that the densification level achievable under SPS is significantly higher than the one obtained by carrying out conventional powder consolidation techniques.

ZrC powder densification mechanisms under SPS conditions were analyzed in the past. Gendre et al. used an empirical model to estimate the stress exponent and the activation energy in SPS of synthesized ZrC_xO_y powder under different loads [23]. This model has been modified recently by Antou et al. with separating intermediate and final sintering stages when investigating the mechanisms contributing to the densification [27]. Wei et al. determined the densification mechanisms of commercial ZrC powder under SPS conditions, in which a densification equation based on the continuum theory of sintering has been used [28]. By carrying out a regression of the obtained equation to the experimental densification data, the strain rate sensitivity and activation energy of the employed ZrC powder were properly assessed [8]. All studies indicated that ZrC exhibits high activation energy and power law creep behavior during the SPS process.

Microstructure coarsening during the final stage of sintering was also observed by Gendre et al., in which the authors attributed this phenomenon to the onset of the pore-grain boundary separation [23]. However, the grain growth mechanism has not been unambiguously identified in that study. Temperature and electric current distributions during SPS of ZrC specimens were also analyzed by a finite element simulation [29]. Despite the fact that porosity of the studied ZrC specimen and the electric contact resistance had not been taken into consideration, a large temperature gradient was identified between the specimen and the SPS tooling area (to which the temperature measuring pyrometer has been focused). This thermal non-uniformity, as stated by the authors, was due to the non-uniform current density distribution in the SPS tooling system as well as the radiative heat loss at the outer surfaces of SPS tooling. It is, therefore, necessary to characterize these thermal effects before analyzing mass transfer and deformation mechanisms in SPS of powder materials.

Both partially and fully dense ZrC products can be utilized for various applications but the respective product service conditions are usually associated with high temperatures. Thermal properties, such as constant pressure specific heat capacity and thermal conductivity of ZrC are, therefore, critical to its potential applications. Measurements conducted a few decades ago on hot-pressed ZrC samples indicated that both heat capacity and thermal conductivity of ZrC increase with temperature [30,31]. However, thermal properties of the SPS processed ZrC have not been reported so far. In addition, the uses of high temperature ceramics sometimes require keeping certain levels of residual porosity in the products (for example, to accommodate volume swelling). In these cases, the specimen's thermal properties largely depend on its relative density because the volume fraction of voids directly determines the amount of substance involved in heat transfer.

In this study, commercial ZrC powders have been subjected to SPS treatments under various processing conditions to produce specimens with a wide range of densities. Conventional hot pressing has also been utilized to consolidate ZrC powder, in which the obtained densification kinetics and microstructures are compared to these retrieved from SPS of ZrC under similar heating and loading profiles. The specimen's temperature is determined using finite element method by correlating the simulated temperature inside the powder specimen with respect to the pyrometer measured temperature at the die surface. The resulting specimen's temperature is utilized to investigate the grain growth mechanism during the final stage of SPS. Both densification and grain growth are studied by hiring recently-developed models [8,32]. The constant pressure specific heat capacity and thermal conductivity of the SPS-processed specimens are measured with respect to temperature, up to 1100 °C. The obtained thermal properties are compared to the reported ones, taking into consideration the relative density level.

2. Materials and Experiment

2.1. Starting Powders

A commercial zirconium (IV) carbide powder (99% metal basis, Sigma-Aldrich Co., St. Louis, MO, USA) was chosen as the tested material in the present study. The as-received powder was first subjected to ultra-sonication (2510 ultra-sonic cleaner, Branson Corp., Danbury, CT, USA) for de-agglomeration. The raw powder was then analyzed by scanning electron microscopy (SEM, Quanta 450, FEI Co., Hillsboro, OR, USA) to examine its morphology. As shown in Figure 1a, a single particle exhibits a polycrystalline structure with inter- and intra-granular pores present. The average grain size of the raw powder is around 1 µm. X-ray diffraction (XRD, X'Pert Pro, PANalytical B.V., Almelo, The Netherlands) of the raw powder was performed using copper as target, diffracted patterns (solid line) are compared to reference peaks (ring markers) along each diffracted plane in Figure 1b. Additionally, the lattice parameter of the starting powder was estimated at every diffracted plane to give an average value of 4.698 Å, which only showed a negligible difference in comparison to the theoretical value (4.699 Å, [31]). The XRD analysis, therefore, has identified the raw powder was very close to the stoichiometry of ZrC.

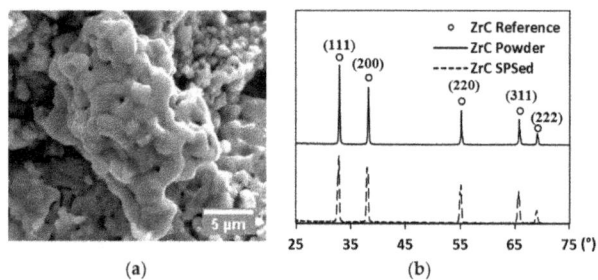

Figure 1. (a) SEM image of the raw powder; and (b) XRD patterns of raw powder (solid line), the SPS-processed specimen (dashed line), and reference peaks (ring markers), respectively.

2.2. Consolidation of Zirconium Carbide Powder

All SPS experiments were performed using a Dr. Sinter SPSS-515 furnace (Fuji Electronic Industrial Co. Ltd., Kawasaki, Japan) with a pulse duration of 3.3 ms and on/off pulse interval of 12:2. For each SPS experiment, 4 g of ZrC powder were used. A 15.3 mm graphite die and two 15 mm graphite punches (I-85 graphite, Electrodes Inc., Santa Fe Springs, CA, USA) had been aligned by inserting well-cut 0.15 mm graphite paper (Fuji Electronic Industrial Co., Ltd., Kawasaki, Japan) in between. The weighted powder was then carefully loaded into the graphite tooling and pre-compacted at room temperature under 3 kN. The geometrical dimensions of a specimen at this point were then used to calculate its green density.

SPS runs were conducted with the maximum processing temperature ranging from 1600 °C to 1800 °C. The following heating profile was used: (i) 6 min from room temperature to 580 °C, 1 min from 580 °C to 600 °C and holding at 600 °C for another 1 minute; (ii) 100 °C/min to 1600 °C and 50 °C/min to target temperature; (iii) dwelling at peak temperature; and (iv) cooling down to 1000 °C and powering off the machine. The temperature was monitored by a digital pyrometer pointing at the lateral surface of the die. The hydraulic uniaxial pressure was consistently applied from the beginning to the end of the consolidation process. The real-time processing parameters, such as temperature, applied load, and axial displacement, were automatically logged by the SPS device.

Hot pressing of the same ZrC powder was carried out using a 50 t hot press furnace (Oxy-Gon Industries, Epsom, NH, USA). The uniaxial pressure was set to 55 MPa. The heating rate was 13 °C/min to 1900 °C. Isothermal holding at 1900 °C was 60 min. In order to make a comparison, "control" SPS runs with same external pressure, heating rate, and holding time were also implemented. By considering the existence of the temperature gap between the specimen and the outer die surface during SPS [33], the peak processing temperature in "control" SPS runs was adjusted to 1600 °C. Such an adjustment aimed at making the actual temperature which the specimen experienced during SPS to be comparable to the one that used in hot pressing (see also Section 3.2). Therefore, the hot pressing and the SPS of ZrC were able to be conducted with imposing similar heating and loading profiles to the powder specimens.

An argon atmosphere was utilized in all SPS and hot-pressing experiments in order to prevent the furnace chamber and the heating elements from being overheated. Graphite tooling was wrapped by carbon felt to reduce heat loss through thermal radiation in SPS runs. For every selected processing profile, an additional run was conducted in the absence of powder. The obtained axial displacement data from this idle run was subtracted from the one retrieved from the real run to provide the true axial shrinkage of a specimen. Every individual experiment was repeated at least twice to ensure the reproducibility of the results.

2.3. Characterization of Processed Specimens

The spark plasma-sintered ZrC specimens have been characterized to reveal their density, open porosity, phase composition, and grain size. All obtained specimens were ground with abrasive SiC paper to remove the adherent graphite foil from their outer surfaces. A specimen's density was first calculated using a geometrical method. If the ratio of the geometrical density of a specimen to the theoretical density of ZrC (6.7 g/cm^3), i.e., the relative density, was more than 90%, the Archimedes method was also applied to reconfirm the obtained value of the relative density. The true axial shrinkage was employed to evaluate the densification kinetics of a specimen with respect to the processing time by assigning a constant radius to the specimen during SPS processing. Open porosity was determined using a helium pycnometer (AccuPyc 1330, Micromeritics Corp., Norcross, GA, USA) by taking into account the difference between apparent and pycnometric relative densities.

After density and open porosity measurements, specimens SPSed at 1700 °C were evenly cut by a precision saw (IsoMet 1000, Buehler, Lake Bluff, IL, USA). The two halves of a specimen were hot-mounted in Bakelite powder with cross-sectional surfaces facing out and subsequently polished with the assistance of a colloidal diamond suspension. Well-polished samples were first analyzed by XRD (X'Pert Pro, PANalytical B.V., Almelo, The Netherlands) to retrieve specimens' phase compositions after SPS consolidation. Then, the polished surfaces were etched for 2 min using HF:HNO$_3$:H$_2$O solution in a volumetric ratio of 1:1:3 in order to have a better reflection of their grain geometries in microstructural characterizations. The obtained micrographs were analyzed by an image software (ImageJ 1.5 g, NIH Image, Bethesda, MD, USA) to calculate the specimen's average grain size based on the mean linear intercept method with a correction factor of 1.5 [34].

2.4. Temperature Evolution in SPS of ZrC

The finite element simulation using COMSOL® Multiphysics software (Comsol Inc., Burlington, MA, USA) was employed to couple electric current and consequent Joule heating phenomena in the implementation of thermal aspects of the employed SPS system. The coupled equations are:

$$\rho_{eff} C_p \frac{\partial T}{\partial t} - \nabla \cdot (k_T \nabla T) = h \tag{1}$$

where ρ_{eff} is the density (kg/m^3); C_p is the heat capacity (J/kg/K) and k_T is the thermal conductivity (W/m/K). h denotes the heat generated by the flowing electric current:

$$h = |J| |E| = \lambda |\nabla V|^2 \tag{2}$$

where J is the electric current density (A/m^2) and E is the intensity of the electric field (V/m); Parameters λ and ∇V correspond to the electric conductivity $\left(\Omega^{-1} \cdot m^{-1}\right)$ and the gradient of electric potential (V/m), respectively. The electric contact resistance between the graphite tooling components was included as:

$$\vec{n} \cdot \vec{J}_{ec} = \frac{1}{R_{ec}} (V_1 - V_2) \tag{3}$$

where $\vec{n}$ is the normal to the contact surface; $\vec{J}_c$ is the generated current density at the contacts (A/m^2); R_{ec} is the electric contact resistance $(\Omega \cdot m^2)$, which has been experimentally derived with respect to the same tooling system [35]; V_1 and V_2 are the electric potential at any two contact surfaces. The effects of thermal contact resistance was implemented by applying the equations developed in [35,36]. The role of horizontal thermal contact resistance was ignored in the simulation as it has been previously determined that its effects on the temperature field are negligible if high pressure is applied [37].

Thermal and electric properties, including the temperature dependence, of the utilized graphite tooling, followed the expressions previously used by Olevsky et al. [32]. ZrC specimen's thermal and electric properties during processing are given in Table 1 as functions of porosity, θ, and temperature, T (K). ZrC's thermal properties were selected in accordance with [15,31].

Table 1. Properties of zirconium carbide used in simulations.

Parameters	Values
Heat capacity, C_p (J/kg/K)	$(352.8 + 0.094T - 2.55 \times 10^3 T^{-2})(1 - \theta)$
Thermal conductivity, k_T (J/m/K)	$\left(\begin{array}{c} 17.82 + 0.024T - 9.39 \times \\ 10^{-6}T^2 + 1.68 \times 10^{-9}T^3 \end{array} \right)(1 - 0.5\theta - 1.5\theta^2)$
Electric conductivity, λ (S/m)	$\frac{1}{39.3 \times 10^{-8} + 76.7 \times 10^{-11}T} \left(\frac{1 - \theta}{1 + 2\theta} \right)$

The SPS machine's logged voltage readings were converted to their root mean square values and interpolated with respect to processing time to provide continuous inputs for the entire modeling process. Figure 2 illustrates the major portion of the tooling-specimen system, which was built as an axial-symmetric model in COMSOL® with specifying the dimensions of each component. During the simulation, the electric potential was introduced at the top electrode (not included in Figure 2), while the bottom one was grounded.

The simulated temperature of the control point at which the temperature measuring pyrometer has been focused was compared to the one obtained from the experiment. These two sets of data have to be in good agreement with each other in order to confirm the reliability of the modeling results and retrieve the specimen temperatures from the simulation. The radial temperature gradient was then calibrated by correlating the calculated control point specimen temperatures with the pyrometer temperatures measured at the control point to allow a suitable comparison of densification kinetics between SPS and hot pressing.

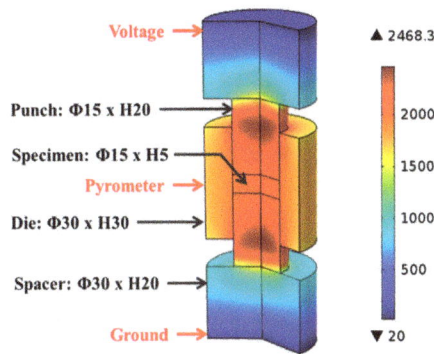

Figure 2. Geometrical model and temperature distribution in finite element simulation, Unit: °C.

2.5. Measurement of Thermal Properties

A series of SPS processed specimens with relative densities ranged from 73.9%–93.3% were further ground to 6 mm diameter by 1 mm thickness disks for thermal property tests. The heat capacity measurements were conducted under constant pressure using the differential scanning calorimeter (DSC 404 F1 Pegasus, Netzsch Co., Selb, Germany) along with the corresponding laser flash apparatus (LFA 427, Netzsch Co., Selb, Germany). The thermal diffusivity was determined by measuring the temperature change on the upper surface of the sample caused by a pulsed laser flash acting on its lower surface. Then, the thermal conductivity was considered to be the product of the sample's heat capacity, density, and its thermal diffusivity calculated by the laser flash apparatus [38]. All tests were performed at every 100 °C interval from room temperature to 1100 °C in an argon atmosphere.

3. Discussion

3.1. Densification Kinetics

The final relative densities of the spark plasma sintered specimens are mapped with processing parameters in Figure 3 (diamond markers). Relative densities of specimens prepared at 1700 °C have been rescaled to be more visible. The density of the hot-pressed specimen (round marker) is also present in comparison to that of the spark plasma sintered one subjected to similar heating and loading profiles (triangle marker). An enhancement in any of the processing parameters leads to an increase of the product's final density. The X-ray diffracted pattern of the SPS-processed specimen is compared to that of the raw powder and to the reference peaks in Figure 1b. The lattice parameter of the SPS specimens was calculated to be ~0.2% larger than that of the raw powder. Such an augmentation might be caused by the free carbon in the raw powder reacting with ZrC during SPS. Since the amount of lattice parameter change could only influence the stoichiometry and lower the theoretical density negligibly, it was considered as minor in the calculation of relative density.

Densification kinetics of spark plasma sintering and hot pressing of ZrC is summarized in Figure 4 with the arrows indicating the onset of the isothermal dwelling. The densification curve of the control SPS appears to possess less data points than that of hot pressing, which is due to the fact that the peak processing temperature in control SPS runs (1600 °C) was intentionally selected to be lower than that in hot pressing (1900 °C). As a result, the hot pressing spent more time to achieve the target temperature. In the control SPS runs, fast densification has already started before the maximum processing temperature has arrived. While in the hot pressing, it is hard to identify the fast densification period until the end of the entire process. Therefore, hot pressing has been evidenced to be much less efficient than SPS in processing ZrC powders.

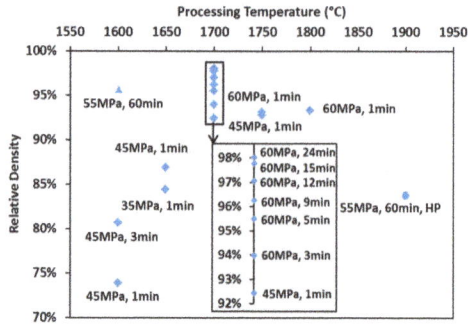

Figure 3. Map of relative densities for specimens prepared under various processing conditions.

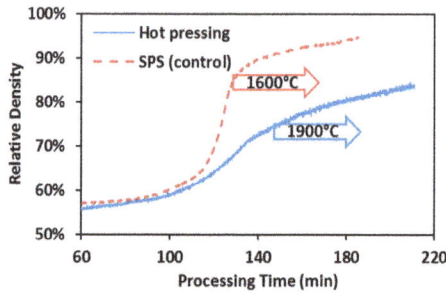

Figure 4. Hot pressing vs. control SPS of ZrC: Comparison of densification kinetics, under 55 MPa and 60 min holding.

Specimens prepared by hot pressing and control SPS processes also gave quite different microstructures when being observed under SEM. As shown in Figure 5, the hot-pressed specimens (Figure 5a) possess a porous structure with visible inter-particle contacts and insignificant signs of grain coarsening. However, under the same magnification, a much more consolidated morphology is present in the SPS-processed specimen (Figure 5b) with clear exhibitions of large grains, while only isolated individual pores are displayed in the matrix.

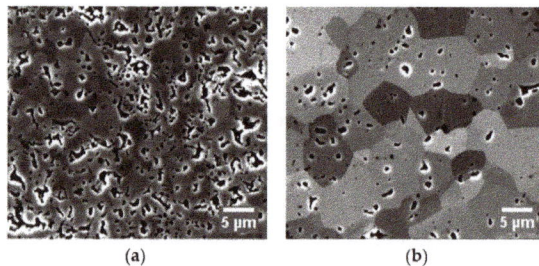

(a)   (b)

Figure 5. Microstructures of ZrC processed by (**a**) hot pressing at 1900 °C; and (**b**) control SPS at 1600 °C under 55 MPa and 60 min holding. The contrast between grains indicates the grain orientations.

3.2. Temperature Evolution in SPS of ZrC

Figure 2 illustrates the temperature distribution obtained from conducting finite element simulation of SPS of ZrC at 1750 °C with color bar indicating the temperature levels on the right.

One can see that the temperature is non-uniformly distributed in the entire system. Simulated temperature values at the point of the pyrometer measurement (long-dash line) and the average temperature in the volume of the specimen (dot-dash line) are plotted in Figure 6, including experimentally-obtained temperature data as a reference (dashed line). The evolution of simulated temperatures at the pyrometer spot show a good agreement with that of the experimental readings, acceptable discrepancies at low temperature range were most likely caused by the lagging of the utilized SPS machine, as well as the radiative heat loss during the rapid heating (100 °C/min) period. However, the specimen's temperatures extracted from the simulation are significantly higher than those retrieved from the experiment and the gaps between these two sets of data keep growing as processing temperature rises. This non-uniform temperature distribution in the tooling system is a common phenomenon in the SPS process and should be carefully assessed [32,39].

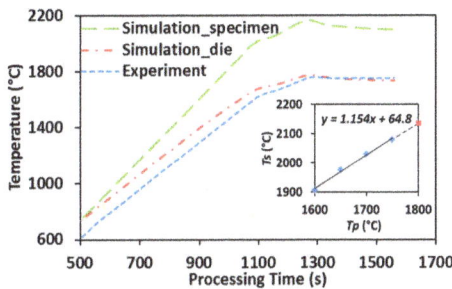

Figure 6. Simulation vs. experiment: temperature evolution in SPS of ZrC (up to 1750 °C).

After plotting the simulated specimen's temperatures (T_s) with respect to the pyrometer-measured processing temperatures (T_p), as shown in the embedded graph of Figure 6, a nearly linear relationship was obtained with processing temperature varying from 1600 °C to 1750 °C. The trend line is similar to the one that has been attained by Antou et al. [29] via finite element simulation, as well as Kelly and Graeve through conducting SPS runs with both top and side pyrometers attached [40]. Additionally, the extrapolation of the obtained relationship has been demonstrated to be able to predict the specimen's temperature when higher SPS temperature is imposed (dashed extension line). Therefore, the temperature experienced by a ZrC specimen subjected to different SPS processing temperatures can be estimated and subsequently used in characterizing densification mechanisms (Section 3.3) and grain growth (Section 3.4).

3.3. Densification Mechanisms in SPS and Hot Pressing of ZrC

In regard to the sintering stages, the hot-pressed ZrC ended up with an 84% relative density which corresponded to the intermediate sintering stage, while the control SPS ZrC has evolved into the final sintering stage with 95% relative density being achieved. Densification mechanisms incorporated in control SPS and hot pressing of ZrC powders under similar heating and loading profiles were investigated to explain the observed different densification kinetics. An analytical/numerical approach for determining the creep coefficients of powder based materials subjected to hot consolidation in a rigid die has been developed recently, in which an analytical densification equation was derived based on the constitutive equation of sintering, as [8]:

$$\dot{\theta} = \frac{d\theta}{dt} = -\frac{A_0}{T}exp\left(-\frac{Q}{RT}\right)(\sigma_z)^{\frac{1}{m}}\left(\frac{3\theta}{2}\right)^{\frac{m+1}{2m}}(1-\theta)^{\frac{m-3}{2m}} \tag{4}$$

where σ_z is the applied axial pressure (*Pa*); T is specimen's absolute temperature (*K*); m is the strain rate sensitivity; Q is the activation energy (*J/mol*); and A_0 is a combined material constant. The creep coefficients, m, Q, and A_0, can be determined through numerically solving Equation (4) in regression to the experimental densification data. A detailed elucidation of such an analysis has been given in [8].

For hot pressing of ZrC, the densification data from the entire isothermal holding stage was selected as the benchmark in regression analysis with the relative density ranging from 75%–84%. At the same time, the selection of densification data from the control SPS runs was taking both the ramping-up and the holding periods into account with relative density increasing from 75%–95%. These selections ensured the same starting porosity (~25%) in both cases. It should be noted that, according to the selected range of relative density, the hot pressing only corresponds to the intermediate sintering stage, while the control SPS includes two sintering stages with the intermediate one preceding the final one [41], and the densification rates associated with these two stages are different (see also Figure 4). Therefore, the study of densification mechanism involved in the intermediate stage was individuated from the one that engaged in the final stage. This approach enabled comparing densification mechanisms incorporated in hot pressing and SPS during the same sintering stage and extended the investigating approach that employed by [8], in which the intermediate and final SPS stages were counted together.

Numerical solutions (Num. soln) are compared to experimental data (Exp. data) in Figure 7. The numerical results are in good agreement with the representative experimental results as shown in Figure 7a, which reveals the reliability of Equation (4) for describing porosity evolution in hot pressing. Porosity evolution during the control SPS has been first split into intermediate (Int) and final stages (Fin) in order to individuate the densification behavior, and then these two stages were put together in one plot (Figure 7b). The discontinuity of the numerical solution at the junction point between the two stages (vertical dot-dash line) reflects the change of creep coefficients.

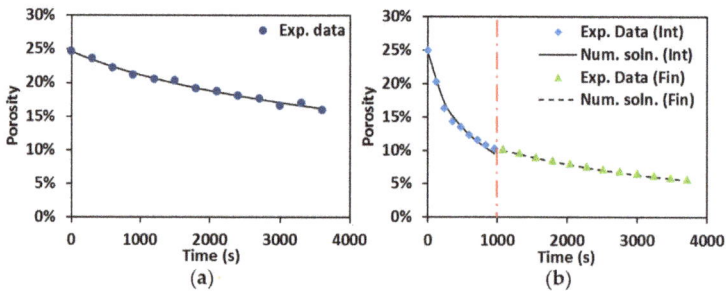

Figure 7. Numerical solution vs. experimental data: (**a**) hot pressing at 1900 °C; and (**b**) control SPS at 1600 °C, under 55 MPa and 60 min holding.

Optimal creep coefficients used in regression analysis are summarized in Table 2 based on the corrected specimen temperature (see Section 3.2). All of the values of the strain rate sensitivity, m, no matter which consolidation technique was used, fall into the range from 0.33 to 0.5. The densification involved in SPS and hot pressing of ZrC is most likely to be grain boundary sliding ($m = 0.5$) associated with dislocation glide ($m = 0.3$) controlled creep [42,43]. Although the m value obtained for the hot pressing was slightly smaller than that of the control SPS obtained for the same sintering stage, the control SPS rendered a significantly lower Q value than the one that the hot pressing provided. Comparatively higher strain rate sensitivity and lower activation energy retrieved from the SPS runs can be attributed to the contribution of electric current, improving the neck growth between particles. Although quantitative evaluations of the current effect in the SPS process are still ongoing [1–3], as shown in [10,11], the inter-particle necks have been observed to be formed at the early SPS stages. Extra atomic diffusional paths created in this manner substantially accelerated the activation of the plastic flow. At the same time, during hot pressing, the inter-particle necks (see Figure 5) appeared to start growing during the intermediate stage and, thus, provided less support for mass transport;

therefore, higher energy was required in the case of hot pressing. The creep coefficients of the control SPS at the intermediate and final sintering stages are nearly identical, except for slightly different values of the activation energies. This difference might be related to the underestimation of the specimen's temperature and the viscous analogue of the shear modulus due to the influences of porosity during the final stage of SPS.

Table 2. Optimal creep coefficients.

Parameters	Intermediate Stage			Final Stage		
	m	*Q* (*kJ/mol*)	A_0	*m*	*Q* (*kJ/mol*)	A_0
Hot pressing	0.382	653	5.92×10^{-6}	N/A	N/A	N/A
Control SPS	0.403	563	6.58×10^{-6}	0.403	576	6.58×10^{-6}

3.4. Grain Growth and Microstructures of SPS Processed Specimens

Average grain sizes (diamond markers) and relative densities (solid line with triangle markers) obtained from specimens produced by SPS processing at 1700 °C are present in Figure 8a with holding time up to 1440 s (24 min). One can see that the increase of the relative densities is accompanied by the augmentation of the grain sizes. Nevertheless, the grain growth appears to be more significant compared to the density evolution. As shown in Figure 8a, the specimens' relative densities range (from 92.3% to 98.1%) indicates the sintering of ZrC has evolved into the final stage when isothermal dwelling started at 1700 °C. During this stage, when the saturation of the temperature level on densification is shown, the processing temperatures still substantially facilitated the grain growth as holding time proceeds [44].

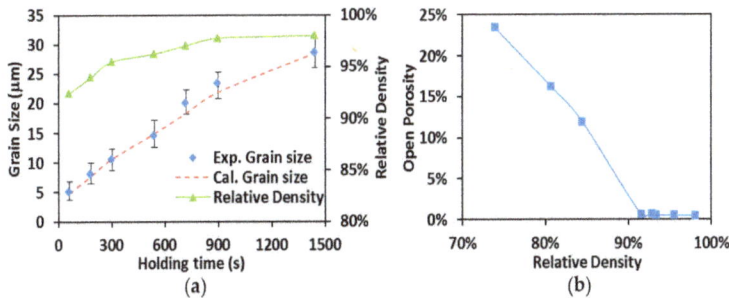

Figure 8. (a) Grain size vs. relative density (SPS at 1700 °C); (b) open porosity vs. relative density.

Chaim stated that, besides temperature and time, the grain growth in SPS of porous ceramics is also controlled by the pore mobility [45]. An equation that includes the dependence of the grain growth on these factors was proposed by Olevsky et al. as [32]:

$$G^p = G_0^p + k_0 t \left(\frac{\theta_c}{\theta + \theta_c} \right)^{\frac{3}{2}} exp \left(-\frac{Q_G}{RT} \right) \tag{5}$$

where G_0 is the initial grain size; p is the grain growth exponent; k_0 is the grain growth constant; θ_c is the critical porosity which reflects the transition from open to close porosity and Q_G is the activation energy for grain growth (*J/mol*).

By using the simulation approach provided in Section 3.2, the specimen's temperature, *T*, was evaluated to be 2303 K (~2030 °C) which is corresponding to a pycnometer-measured temperature of 1700 °C. The critical porosity, θ_c, was determined through the open porosity measurements. The specimen's open porosities are plotted with respect to their relative densities in Figure 8b. The decrease of open porosity suddenly turns into a plateau with open porosity close to zero after relative density reaches 93%, indicating the open pores in these specimens are nearly gone. The turning point Figure 8b was, therefore, considered to be the moment of transition from open porosity to close porosity and the value of θ_c was set to 0.07 in the evaluation of other grain growth coefficients.

An Excel® Solver program (Microsoft, Redmond, WA, USA) was used to assess the values of p, k_0, and Q_G. By iteratively optimizing these values, as demonstrated by the dashed line in Figure 8a, Equation (5) produced a set of calculated grain sizes which consistently agree with the ones obtained from the experiments. Additionally, the coefficient optimization gave a grain growth exponent of $p \approx 2$, which corresponds to the grain boundary diffusion controlled grain growth [45]. The observed insignificant change in density during the final holding stage in the present study is in agreement with the study of Djohari et al., in which the grain boundary diffusion has been described as a cause of virtually little densification in the later stage of sintering [46]. Furthermore, the activation energy for grain growth was estimated to be 290 kJ/mol. This value is way lower than the activation energies found for zirconium lattice diffusion (720 kJ/mol, [47]), for carbon bulk self-diffusion (470 kJ/mol, [48]) in ZrC_x and for creep-introduced densification (576 kJ/mol, see also Table 2), suggesting that the grain growth was preferred during the final stage of SPS of ZrC compared to other mechanisms.

The representative micrographs of specimens' cross-sectional surfaces are illustrated in Figure 9, from where a direct impression of how grains interact with inter-granular pores at the triple junctions can be obtained: the grain growth gradually contributes to the process of pore closure. It appears that the densification can benefit from the grain growth to a certain degree in the final stage. However, this phenomenological observation could be complemented by nano- or atomic- scale analyses to reveal the actual mass transfer mechanism (motions of grains or dislocations). The existence of the amount of intra-granular pores in the microstructures of all specimens is possibly due to: (i) internal pores from initial powder (see also Figure 1a); (ii) high-temperature pore formation mechanisms proposed by Kelly and Graeve [40]. The contrast between grains indicates the grain orientations. The contrast difference seems to become significant with increasing holding time suggesting that the grain growth was associated with the grain movements.

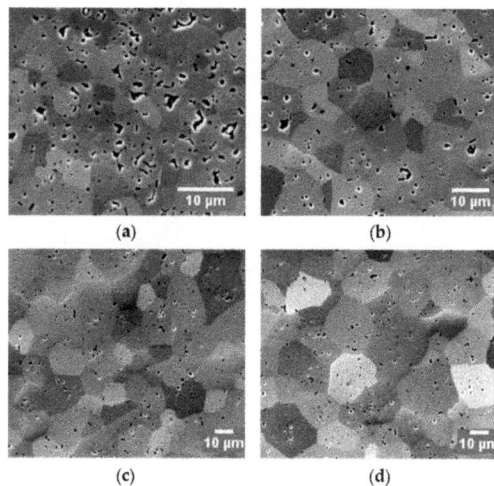

(a)   (b)

(c)   (d)

Figure 9. Microstructures of SPS-processed specimens at 1700 °C with: (**a**) 1 min; (**b**) 9 min; (**c**) 15 min; and (**d**) 24 min holding time, all under a pressure of 60 MPa.

3.5. Thermal Properties of SPS-Processed Specimens

The heat capacity of specimens SPSed under various processing conditions increase with elevating temperature, as well as with raising the relative density (see Figure 10). Heat capacity first rises rapidly from room temperature to 300 °C, and then it grows slowly until 1100 °C. According to [49], the Debye temperature of stoichiometric ZrC is between 500 and 600 K (200~300 °C), suggesting that the observation from the present study is in accordance with the reported data, as the heat capacity of carbide at low temperatures depends on its Debye temperature. Additionally, for a given volume, a specimen with higher relative density possesses more thermal mass, therefore, more heat is required for a degree of temperature rise. Heat capacities of fully-dense ZrC were extrapolated from the measurements of partially-dense specimens and compared with those calculated by Turchanin et al. using both Debye and Einstein equations [50] in the same graph. It shows that the highest heat capacity obtained from this study is very close to the one reported in the past while the extrapolation is more accurate as the temperature goes over 200 °C.

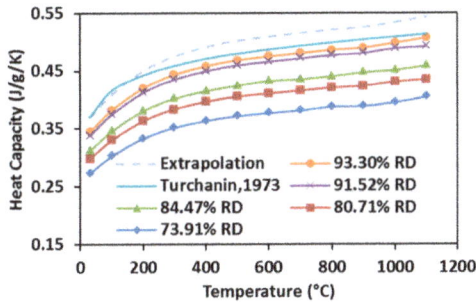

Figure 10. Heat capacities of SPS-processed specimens as a function of temperature.

As shown in Figure 11, the thermal conductivities of SPS processed specimens rise with increasing temperature in the tested temperature range. This observation indicates quite unique ZrC properties compared to many other ceramic materials and it has been primarily attributed to the contributions of conduction electron bands and high phonon conductivity in ceramics materials [15]. Additionally, the thermal conductivity is shown to increase with enhancing the relative density because higher relative density is associated with the presence of fewer pores, hence more thermal pathways are present in the processed specimen. Thermal conductivities of the hot-pressed ZrC with very similar relative density (~93.3%) obtained by Taylor were considered to be the highest results that have been reported in the past [30]. These data have been included for comparison in Figure 11 (scatter diamond markers, no data reported for temperature below 600 °C). It appears that the measured thermal conductivities from the SPS-processed specimens are higher than those from the hot-pressed ones. Although the method of characterization between the present study and [30] is very different, the obtained evolutions of thermal conductivities are consistent and the flash method appears to be able to retrieve them at lower temperatures in a shorter time.

SPS-processed specimens exhibited excellent heat capacities and thermal conductivities compared to these reported in the past. The improvements of the thermal properties are most likely due to the reduction of impurities during the SPS process. Impurities are easy to be introduced into powders during manufacturing processes since powders have large surface area and high surface energy. The impurities or secondary atoms usually occupy lattice vacancies or present as interstitials which act as strong scattering centers for phonons and electrons. These impurities are hard to remove during conventional sintering processes. Therefore, both thermal and electrical properties of the sintered product can be negatively influenced. The SPS process provides high electric current enabling the generation of micro-discharges along powder surfaces to remove impurities [51,52] and, in turn, to improve the above-mentioned properties of the final products.

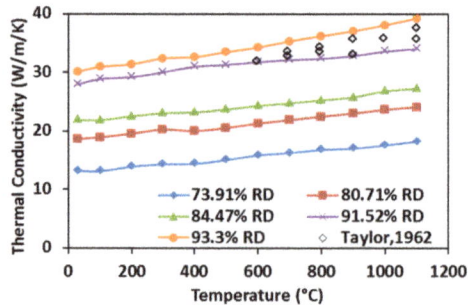

Figure 11. Thermal conductivities of SPS-processed specimens as a function of temperature.

4. Conclusions

ZrC pellets with high relative densities have been successfully produced by SPS. Relative densities of obtained specimens were mapped with processing temperature, applied pressure, and holding time to elucidate the effects of these processing parameters on the densification level. Hot pressing and SPS of ZrC were carried out in the conducted comparative study to investigate the different densification mechanisms affecting these two techniques. Higher strain rate sensitivity and lower activation energy are observed for the control SPS compared to those observed for the conventional hot pressing. The causes of these differences have been attributed to the effects of the electric current during SPS processing.

Temperature evolution during SPS of ZrC was implemented by a finite element simulation to characterize the thermal gradient between the die surface and the specimen. The specimen's actual temperature was verified by correlating the simulated temperatures with respect to the pyrometer measured ones. The specimen's temperature was then substituted into recently modified models to study the grain growth kinetics in the final stage of SPS, and the grain boundary diffusion was determined to be the major control mechanism. The microscopic examinations of specimen's cross-sectional area also reflected that the grain growth in the final SPS stage contributes to the closure of the inter-granular pores.

Specific heat capacities and thermal conductivities of the SPS processed specimens were measured from room temperature to 1100 °C using DSC along with LFA. Specimens' thermal properties were found to increase either with higher relative density or with raising temperature. The thermal properties obtained from the SPS-processed specimens were higher than the reported data retrieved from the hot-pressed samples at the similar relative density level, thereby indicating the impurity cleaning effect during the SPS process.

Both experimental and modeling approaches have been conducted to characterize the hot consolidation of ZrC. The obtained results can be used for future optimization purposes, including the possible design of material structures in a sophisticated way.

Acknowledgments: The support of the U.S. Department of Energy, Materials Sciences Division, under Award No. DE-SC0008581 is gratefully acknowledged. The authors also acknowledge the assistance of Steve Barlow, and the use of SEM equipment at the San Diego State University Electron Microscopy Facility acquired by NSF instrumentation grant DBI-0959908.

Author Contributions: Xialu Wei, Christina Back, Oleg Izhvanov, Christopher D. Haines and Eugene A. Olevsky conceived and designed the experiments; Xialu Wei performed the experiments; Xialu Wei, Christopher D. Haines and Eugene A. Olevsky. analyzed the data; Oleg Izhvanov and Christina Back contributed materials; Xialu Wei wrote the paper.

Conflicts of Interest: The authors declare no conflict of interest.

References

1. Munir, Z.A.; Anselmi-Tamburini, U.; Ohyanagi, M. The effect of electric field and pressure on the synthesis and consolidation of materials: A review of the spark plasma sintering method. *J. Mater. Sci.* **2006**, *41*, 763–777. [CrossRef]
2. Orru, R.; Licheri, R.; Locci, A.M.; Cincotti, A.; Cao, G.C. Consolidation/synthesis of materials by electric current activated/assisted sintering. *Mater. Sci. Eng. R.* **2009**, *63*, 127–287. [CrossRef]
3. Munir, Z.A.; Quach, D.V.; Ohyanagi, M. Electric current activation of sintering: A review of the pulsed electric current sintering process. *J. Am. Ceram. Soc.* **2011**, *94*, 1–19. [CrossRef]
4. Guillon, O.; Gonzalez-Julian, J.; Dargatz, B.; Kessel, T.; Schierning, G.; Räthel, J.; Herrmann, M. Field-assisted sintering technology/spark plasma sintering: Mechanisms, materials, and technology developments. *Adv. Eng. Mater.* **2014**, *16*, 830–849. [CrossRef]
5. Khaleghi, E.; Lin, Y.S.; Meyers, M.A.; Olevsky, E.A. Spark plasma sintering of tantalum carbide. *Scr. Mater.* **2010**, *63*, 577–580. [CrossRef]
6. Bellosi, A.; Monteverde, F.; Sciti, D. Fast Densification of ultra-high-temperature ceramics by spark plasma sintering. *Int. J. Appl. Ceram. Technol.* **2006**, *3*, 32–40. [CrossRef]
7. Li, W.; Olevsky, E.A.; Khasanov, O.L.; Back, C.A.; Izhvanov, O.L.; Opperman, J.; Khalifa, H.E. Spark plasma sintering of agglomerated vanadium carbide powder. *Ceram. Int.* **2015**, *41*, 3748–3759. [CrossRef]
8. Wei, X.; Back, C.; Izhvanov, O.; Khasanov, O.; Haines, C.; Olevsky, E. Spark plasma sintering of commercial zirconium carbide powders: Densification behavior and mechanical properties. *Materials* **2015**, *8*, 6043–6061. [CrossRef]
9. Mizuguchi, T.; Guo, S.; Kagawa, Y. Transmission electron microscopy characterization of spark plasma sintered ZrB2 ceramic. *Ceram. Int.* **2010**, *36*, 943–946. [CrossRef]
10. Giuntini, D.; Wei, X.; Maximenko, A.L.; Li, W.; Ilyina, A.M.; Olevsky, E.A. Initial stage of free pressureless spark-plasma sintering of vanadium carbide: Determination of surface diffusion parameters. *Int. J. Refract. Met. Hard Mater.* **2013**, *41*, 501–506. [CrossRef]
11. Olevsky, E.; Bogachev, I.; Maximenko, A. Spark-plasma sintering efficiency control by inter-particle contact area growth: A viewpoint. *Scr. Mater.* **2013**, *69*, 112–116. [CrossRef]
12. Narayan, J. Grain growth model for electric field-assisted processing and flash sintering of materials. *Scr. Mater.* **2013**, *68*, 785–788. [CrossRef]
13. Ryu, H.J.; Lee, Y.W.; Cha, S.I.; Hong, S.H. Sintering behaviour and microstructures of carbides and nitrides for the inert matrix fuel by spark plasma sintering. *J. Nucl. Mater.* **2006**, *352*, 341–348. [CrossRef]
14. Vasudevamurthy, G.; Knight, T.W.; Roberts, E.; Adams, T.M. Laboratory production of zirconium carbide compacts for use in inert matrix fuels. *J. Nucl. Mater.* **2008**, *374*, 241–247. [CrossRef]
15. Katoh, Y.; Vasudevamurthy, G.; Nozawa, T.; Snead, L.L. Properties of zirconium carbide for nuclear fuel applications. *J. Nucl. Mater.* **2013**, *441*, 718–742. [CrossRef]
16. Spivak, I.I.; Klimenko, V.V. Densification kinetics in the hot pressing and recrystallization of carbides. *Sov. Powder Metall. Met. Ceram.* **1973**, *12*, 883–887.
17. Bulychev, V.P.; Andrievskii, R.A.; Nezhevenko, L.B. The sintering of zirconium carbide. *Sov. Powder Metall. Met. Ceram.* **1977**, *16*, 273–276. [CrossRef]
18. Barnier, P.; Brodhag, C.; Thevenot, F. Hot-pressing kinetics of zirconium carbide. *J. Mater. Sci.* **1986**, *21*, 2547–2552. [CrossRef]
19. Minhaga, E.; Scott, W.D. Sintering and mechanical-properties of ZrC-ZrO2 composites. *J. Mater. Sci.* **1988**, *23*, 2865–2870. [CrossRef]
20. Wang, X.G.; Guo, W.M.; Kan, Y.M.; Zhang, G.J.; Wang, P.L. Densification behavior and properties of hot-pressed ZrC ceramics with Zr and graphite additives. *J. Eur. Ceram. Soc.* **2011**, *31*, 1103–1111. [CrossRef]
21. Zhao, L.; Jia, D.; Duan, X.; Yang, X.; Zhou, Y. Pressureless sintering of ZrC-based ceramics by enhancing powder sinterability. *Int. J. Refract. Met. Hard Mater.* **2011**, *29*, 516–521. [CrossRef]
22. Sciti, D.; Guicciardi, S.; Nygren, M. Spark plasma sintering and mechanical behaviour of ZrC-based composites. *Scr. Mater.* **2008**, *59*, 638–641. [CrossRef]

23. Gendre, M.; Maître, A.; Trolliard, G. A study of the densification mechanisms during spark plasma sintering of zirconium (oxy-)carbide powders. *Acta Mater.* **2010**, *58*, 2598–2609. [CrossRef]
24. Gendre, M.; Maître, A.; Trolliard, G. Synthesis of zirconium oxycarbide (ZrC$_x$O$_y$) powders: Influence of stoichiometry on densification kinetics during spark plasma sintering and on mechanical properties. *J. Eur. Ceram. Soc.* **2011**, *31*, 2377–2385. [CrossRef]
25. Núñez-González, B.; Ortiz, A.L.; Guiberteau, F.; Nygren, M.; Shaw, L. Improvement of the spark-plasma-sintering kinetics of ZrC by high-energy call-milling. *J. Am. Ceram. Soc.* **2012**, *95*, 453–456. [CrossRef]
26. Xie, J.; Fu, Z.; Wang, Y.; Lee, W.W.; Niihara, K. Synthesis of nanosized zirconium carbide powders by a combinational method of sol–gel and pulse current heating. *J. Eur. Ceram. Soc.* **2014**, *34*, 13.e1–13.e7. [CrossRef]
27. Antou, G.; Pradeilles, N.; Gendre, M.; Maître, A. New approach of the evolution of densification mechanisms during Spark Plasma Sintering: Application to zirconium (oxy-)carbide ceramics. *Scr. Mater.* **2015**, *101*, 103–106. [CrossRef]
28. Olevsky, E.A. Theory of sintering: From discrete to continuum. *Mater. Sci. Eng. R.* **1998**, *23*, 41–100. [CrossRef]
29. Antou, G.; Mathieu, G.; Trolliard, G.; Maître, A. Spark plasma sintering of zirconium carbide and oxycarbide: Finite element modeling of current density, temperature, and stress distributions. *J. Mater. Res.* **2009**, *24*, 404–412. [CrossRef]
30. Taylor, R.E. Thermal conductivity of zirconium carbide at high temperatures. *J. Am. Ceram. Soc.* **1962**, *45*, 353–354. [CrossRef]
31. Grossman, L.N. High-temperature thermophysical properties of zirconium carbide. *J. Am. Ceram. Soc.* **1965**, *48*, 236–242. [CrossRef]
32. Olevsky, E.A.; Garcia-Cardona, C.; Bradbury, W.L.; Haines, C.D.; Martin, D.G.; Kapoor, D. Fundamental aspects of spark plasma sintering: II. Finite element analysis of scalability. *J. Am. Ceram. Soc.* **2012**, *95*, 2414–2422. [CrossRef]
33. Giuntini, D.; Olevsky, E.A.; Garcia-Cardona, C.; Maximenko, A.L.; Yurlova, M.S.; Haines, C.D. Localized overheating phenomena and optimization of spark-plasma sintering tooling design. *Materials* **2013**, *6*, 2612–2632. [CrossRef]
34. Meyers, M.A.; Chawla, K.K. *Mechanical Behavior of Materials*; Cambridge University Press: Cambridge, UK, 2009.
35. Wei, X.; Giuntini, D.; Maximenko, A.L.; Haines, C.D.; Olevsky, E.A. Experimental investigation of electric contact resistance in spark plasma sintering tooling setup. *J. Am. Ceram. Soc.* **2015**, *98*, 3553–3560. [CrossRef]
36. Yovanovich, M.M. Four decades of research on thermal contact, gap and joint resistance in microelectronics. *IEEE. Trans. Compon. Pack. Technol.* **2005**, *28*, 182–206. [CrossRef]
37. Anselmi-Tamburini, U.; Gennari, S.; Garay, J.E.; Munir, Z.A. Fundamental investigations on the spark plasma sintering/synthesis process: II. Modeling of current and temperature distributions. *Mater. Sci. Eng. A* **2005**, *394*, 139–148. [CrossRef]
38. Parker, W.J.; Jenkins, R.J.; Butler, C.P.; Abbott, G.L. Flash method of determining thermal diffusivity, heat capacity, and thermal conductivity. *J. Appl. Phys.* **1961**, *32*, 1679–1684. [CrossRef]
39. Giuntini, D.; Raethel, J.; Herrmann, M.; Michaelis, A.; Olevsky, E.A. Advancement of tooling for spark plasma sintering. *J. Am. Ceram. Soc.* **2015**, *98*, 3529–3537. [CrossRef]
40. Kelly, J.P.; Graeve, O.A. Mechanisms of pore formation in high-temperature carbides: Case study of TaC prepared by spark plasma sintering. *Acta Mater.* **2015**, *84*, 472–483. [CrossRef]
41. Coble, R.L. Sintering crystalline solids. I. Intermediate and final state diffusion models. *J. Appl. Phys.* **1961**, *32*, 787–792. [CrossRef]
42. Gifkins, R.C. Grain-boundary sliding and its accommodation during creep and superplasticity. *Metall. Trans. A* **1976**, *7*, 1225–1232. [CrossRef]
43. Weertman, J. Steady-state creep of crystals. *J. Appl. Phys.* **1957**, *28*, 1185–1189. [CrossRef]
44. Chen, I.W.; Wang, X.H. Sintering dense nanocrystalline ceramics without final-stage grain growth. *Nature* **2000**, *404*, 168–171. [CrossRef] [PubMed]
45. Chaim, R. Densification mechanisms in spark plasma sintering of nanocrystalline ceramics. *Mater. Sci. Eng. A* **2007**, *443*, 25–32. [CrossRef]
46. Djohari, H.; Martínez-Herrera, J.I.; Derby, J.J. Transport mechanisms and densification during sintering: I. Viscous flow versus vacancy diffusion. *Chem. Eng. Sci.* **2009**, *64*, 3799–3809. [CrossRef]

47. Andrievskii, R.A.; Khormov, Y.F.; Alekseeva, I.S. Self-diffusion of carbon and metal atoms in zirconium and niobium carbides (in Russian). *Fiz. Metal. Metalloved.* **1971**, *32*, 664–667.
48. Sarian, S. Diffusion of carbon through zirconium monocarbide. *J. Appl. Phys.* **1967**, *38*, 1794–1798. [CrossRef]
49. Storms, E.K. *The Refractory Carbides*; Academic Press: New York, NY, USA, 1967.
50. Turchanin, A.G.; Guseva, E.A.; Fesenko, V.V. Thermodynamic properties of refractory carbides in the temperature range 0–3000° K. *Sov. Powder Metall. Met. Ceram.* **1973**, *12*, 215–217. [CrossRef]
51. Anderson, K.R.; Groza, J.R.; Fendorf, M.; Echer, C.J. Surface oxide debonding in field assisted powder sintering. *Mater. Sci. Eng. A* **1999**, *270*, 278–282. [CrossRef]
52. Groza, J.R.; Zavaliangos, A. Sintering activation by external electrical field. *Mater. Sci. Eng. A* **2000**, *287*, 171–177. [CrossRef]

materials

MDPI

Article

Processing, Mechanical and Optical Properties of Additive-Free ZrC Ceramics Prepared by Spark Plasma Sintering

Clara Musa [1], Roberta Licheri [1], Roberto Orrù [1,*], Giacomo Cao [1], Diletta Sciti [2], Laura Silvestroni [2], Luca Zoli [2], Andrea Balbo [2,3], Luca Mercatelli [4], Marco Meucci [4] and Elisa Sani [4]

[1] Dipartimento di Ingegneria Meccanica, Chimica e dei Materiali, Unità di Ricerca del Consorzio Interuniversitario Nazionale per la Scienza e Tecnologia dei Materiali (INSTM)—Università degli Studi di Cagliari, via Marengo 2, Cagliari 09123, Italy; claramusa80@gmail.com (C.M.); roberta.licheri@dimcm.unica.it (R.L.); giacomo.cao@dimcm.unica.it (G.C.)

[2] ISTEC-CNR, Institute of Science and Technology for Ceramics, Via Granarolo 64, Faenza 48018, Italy; diletta.sciti@istec.cnr.it (D.S.); laura.silvestroni@istec.cnr.it (L.S.); luca.zoli@istec.cnr.it (L.Z.); andrea.balbo@unife.it (A.B.)

[3] Corrosion and Metallurgy Study Centre "Aldo Daccò", Engineering Department, University of Ferrara, G. Saragat 4a, Ferrara 44122, Italy

[4] INO-CNR, National Institute of Optics, Largo E. Fermi, 6, Firenze 50125, Italy; luca.mercatelli@ino.it (L.M.); marco.meucci@ino.it (M.M.); elisa.sani@ino.it (E.S.)

* Correspondence: roberto.orru@dimcm.unica.it; Tel.: +39-070-6755076; Fax: +39-070-6755057

Academic Editor: Eugene A. Olevsky
Received: 21 April 2016; Accepted: 9 June 2016; Published: 18 June 2016

Abstract: In the present study, nearly fully dense monolithic ZrC samples are produced and broadly characterized from microstructural, mechanical and optical points of view. Specifically, 98% dense products are obtained by Spark Plasma Sintering (SPS) after 20 min dwell time at 1850 °C starting from powders preliminarily prepared by Self-propagating High-temperature Synthesis (SHS) followed by 20 min ball milling. A prolonged mechanical treatment up to 2 h of SHS powders does not lead to appreciable benefits. Vickers hardness of the resulting samples (17.5 ± 0.4 GPa) is reasonably good for monolithic ceramics, but the mechanical strength (about 250 MPa up to 1000 °C) could be further improved by suitable optimization of the starting powder characteristics. The very smoothly polished ZrC specimen subjected to optical measurements displays high absorption in the visible-near infrared region and low thermal emittance at longer wavelengths. Moreover, the sample exhibits goodspectral selectivity (2.1–2.4) in the 1000–1400 K temperature range. These preliminary results suggest that ZrC ceramics produced through the two-step SHS/SPS processing route can be considered as attractive reference materials for the development of innovative solar energy absorbers.

Keywords: Spark Plasma Sintering; self-propagating high-temperature synthesis; ultra-high-temperature-ceramics; carbides; mechanical properties; optical properties

1. Introduction

Due to the peculiar combination of its chemico-physical and mechanical properties, such as high melting temperature (above 3500 °C), hardness, low density, chemical inertness, good electrical and thermal conductivity, zirconium carbide (ZrC) has been acknowledged as a very promising material for high temperature applications [1]. Moreover, ZrC displays low neutron absorption and selective solar energy absorption, which makes it particularly attractive in the nuclear [2] and solar energy [3] fields, respectively.

Various synthesis routes are currently available for the preparation of ZrC powders [4–12]. These include the carbo-thermal reduction of zirconia in high temperature furnaces [4], mechano-chemistry [5], solution methods [6], sol-gel [8], and self-propagating high-temperature synthesis (SHS) [7,9–12].

In spite of the availability of the various synthesis options reported above, the strong covalent Zr-C bond makes the fabrication of dense monolithic ZrC bodies a difficult achievement. For instance, it was reported that the pressureless sintering of pure ZrC powders for 60 min at 1950 °C leads to extremely porous samples (about 70% relative density) [13]. Final density was increased up to 94.4% only upon sintering at 2100 °C for 2 h [14]. The densification behavior is generally improved in the presence of mechanical loads, although the theoretical density is hard to be reached when considering classical hot pressing methods [15,16]. For example, 91% of dense samples are produced after 1 h of holding time at 2000 °C and 30 MPa applied pressure [16]. Modest densification levels are also reached when the reaction synthesis and densification of pure ZrC was performed in one single processing stage by reactive hot-pressing [14]. In contrast, significant improvements can be obtained when considering the Spark Plasma Sintering (SPS) technology, an efficient consolidation method where powder compacts are rapidly heated by the electric current flowing through the conductive die containing them [17]. For this reason, various studies involving the use of the SPS method for the densification of ZrC powders were carried out in the last decade [18–23]. Specifically, bulk products with relative density above 97% are generally obtained from as-received commercial ZrC powder by SPS when operating at temperatures of 2100 °C or higher values [18,20,23]. A mechanical treatment of the starting powders was reported to promote their densification by SPS [20,22]. Moreover, about 97.9% dense ZrC samples were recently obtained by SPS at 1800 °C, when the applied pressure was increased to 200 MPa, which was made possible by the use of a specifically designed double-die configuration [23].

Likewise for other members of ultra-high temperature ceramics (UHTCs), the consolidation of ZrC powders can be also made easier by the introduction of appropriate sintering aids, although the presence of secondary phases might not be desirable for certain high-temperature applications.

The difficulties encountered for the fabrication of fully dense ZrC products are also responsible for the lack of reliable data for related key properties, particularly for the thermo-mechanical and optical ones, which, on the other hand, are necessary to define the possible exploitation of the material in high-temperature solar absorbers.

This study deals with the fabrication of dense monophasic zirconium carbide by combining the SHS and SPS techniques. Specifically, according to previous findings providing evidence of the improved sintering ability of combustion synthesized powders with respect to differently prepared products [24,25], the zirconium carbide phase is first obtained by SHS. A systematic investigation is then performed to identify the optimal SPS temperature and time conditions to obtain high densification levels. The resulting bulk products are finally characterized from the microstructural, thermo-mechanical and optical points of view.

2. Materials and Methods

Zirconium (Alfa Aesar, product code 00418, particle size < 44 µm, purity > 98.5%, Karlsruhe, Germany) and graphite (Sigma-Aldrich, product code 282863, particle size < 20 µm, purity > 99.99%, St. Louis, MO, USA) powders were used as starting reactants for the synthesis of ZrC by SHS. Mixing of reagents was performed in agreement with the following reaction stoichiometry:

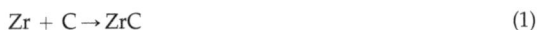

$$Zr + C \rightarrow ZrC \qquad (1)$$

About 15 g of the elemental powders were mixed for 20 min by means of a SPEX 8000 shaker mill (SPEX CertiPrep, Metuchen, NJ, USA) using plastic vials and 6 zirconia balls (2 mm diameter, 0.35 g) (Union process, Akron, OH, USA). SHS experiments were then conducted on cylindrical pellets (10 mm diameter, 20–30 mm height) prepared by uni-axially pressing 8–10 g of the obtained mixture.

The synthesis process was carried out inside a reaction chamber under Ar atmosphere. An electrically heated tungsten wire (Inland Europe, Varigney, France) was used to locally activate the SHS reaction. The combustion temperature during synthesis evolution was measured using a two-color pyrometer (Ircon Mirage OR 15-990, Santa Cruz, CA, USA) focused on the center of the lateral surface of reacting samples. Combustion front velocity was estimated on the basis of the frame-by-frame analysis of the video recording. Further details relative to SHS experiments can be found elsewhere [26].

The SHS products were converted to powder form by subsequent ball milling (SPEX CertiPrep, Metuchen, NJ, USA). Specifically, about 4 g of the obtained porous samples have been mechanically treated using a stainless steel vial (VWR International PBI, Milan, Italy) with two steel balls (13 mm diameter, 4 g) to obtain a ball-to-powder weight, or charge ratio (CR), equal to two. The milling time (t_M) was varied in the range 5–120 min. Particle size of the resulting powders was determined using a laser light scattering analyzer (CILAS 1180, Orleans, France).

Iron contamination from milling media to ZrC powders was evaluated by means of Inductively Coupled Plasma Optical Emission Spectroscopy (ICP-OES, Optima 5300DV Perkin Elmer, Waltham, MA, USA). The complete powder dissolution for this analysis was obtained using a hot mixture of nitric acid and hydrochloric acid (Carlo Erba Reagents, Milan, Italy) in a molar ratio of 1:3 (*aqua regia*).

An SPS equipment (515 model, Sumitomo Coal Mining Co., Ltd., Kanagawa, Japan) was used under vacuum (20 Pa) conditions for consolidation of the differently ball-milled SHS powders. This apparatus basically consists of a DC pulsed current generator (10 V, 1500 A, 300 Hz) (Sumitomo Coal Mining Co., Ltd., Kanagawa, Japan) combined with a uniaxial press (max 50 kN). The pulse cycle of the SPS machine was set to 12 ms on and 2 ms off, with a characteristic time of each pulse equal to 3.3 ms. About 3.6 g of powders were first cold-compacted inside the die (30 mm outside diameter; 15 mm inside diameter; 30 mm height) to produce 14.7 mm diameter specimens. To facilitate sample release after sintering, the internal die surface was previously lined with graphite foils (0.13 mm thick, Alfa Aesar, Karlsruhe, Germany). In order to control the evolution of the SPS process, temperature, current intensity, voltage, and sample displacement were monitored and recorded in real time. In particular, temperature was measured using a C-type thermocouple (W-Re, 250 μm diameter, Fuji Electronic Industrial Co., Ltd., Kanagawa, Japan) inserted in a small hole drilled on the die. SPS experiments were initiated with the imposition of a prescribed thermal cycle, where the temperature was first increased from the room value to the maximum level (T_D) in 10 min (t_H). Then, the T_D value was kept constant for a prescribed duration (t_D). The effect of T_D and t_D on the density of the sintered product was investigated in the 1750–1850 °C and 5–20 min ranges, respectively. The mechanical pressure of 50 MPa was applied from the beginning of each SPS experiment. For the sake of reproducibility, each experiment was repeated at least twice. Additional detailed information on SPS experiments is reported in previous studies [27,28].

In addition to the standard specimens (14.7 mm diameter), larger size samples were prepared for the mechanical characterization, in particular for the evaluation of flexural strength properties. However, significantly higher current levels with respect to those ones obtainable with the SPS 515 equipment (Sumitomo Coal Mining Co., Ltd., Kanagawa, Japan) (1500 A at most) are required to densify such larger samples. Thus, another SPS apparatus, namely the HPD 25-1 model (FCT Systeme GmbH, Rauenstein, Germany), able to provide current intensities up to 8000 A, was used to produce 40 mm diameter samples. In this regard, it should be noted that, for the latter equipment, temperature is measured and controlled at an axial point, whereas, as specified above, in the Sumitomo apparatus, such measurements are made on the surface of the die.

Relative densities of the sintered specimens were determined by the Archimedes' method using distilled water as immersion medium and by considering the theoretical value for ZrC equal to 6.73 g/cm^3 [29].

The crystalline phases were identified using an X-ray diffractometer (Philips PW 1830, Almelo, The Netherlands) equipped with a Ni filtered Cu K$_\alpha$ radiation (λ = 1.5405 Å). A Rietveld analytical procedure was employed to evaluate the average crystallite size [30]. The morphology of SHSed powders, as well as the microstructure and local phase composition of the sintered samples, were examined by scanning electron microscopy (SEM) (mod. S4000, Hitachi, Tokyo, Japan and mod. ΣIGMA, ZEISS NTS Gmbh, Oberkochen, Germany) and energy dispersive X-rays spectroscopy (EDS) (Kevex Sigma 32, Noran Instruments, Middleton, WI, USA and mod. INCA Energy 300, Oxford Instruments, Abingdon, UK), respectively.

Vickers microhardness (HV1.0) was measured with a load of 9.81 N, using a Zwick 3212 tester (Zwick GmbH & Co., Ulm, Germany), according to the European standard prEN 843-4.

With the same Zwick 3212 tester, the fracture toughness (K_{IC}) was evaluated by the direct crack measurement method on the polished surfaces with a load of 98.1 N, by considering a radial-median crack and using the equation of Evans and Charles [31].

The 4-pt flexural strength (σ) was measured at room temperature (RT) and up to 1200 °C in partially protective Ar atmosphere, using the guidelines of the European standards for advanced ceramics ENV843-1:2004 [32] and EN820-1:2002 [33], respectively. Chamfered type-A bars with dimensions $25.0 \times 2.5 \times 2.0$ mm^3 (length by width by thickness, respectively) were tested at RT using a semi-articulated steel-made 4-pt fixture (lower span 20 mm, upper span 10 mm) in a screw-driven load frame (Zwick-Roell mod. Z050, Ulm, Germany), 1 mm/min of cross-head speed. Flexural strength at a high temperature was instead measured in a partially protective Ar environment using an adapted furnace (mod. HTTF, Severn Furnaces Ltd., Draycott Business Park, Durseley, UK) mounted on an Instron apparatus (mod. 6025) (Instron, Illinois Tool Works Inc., Norwood, MA, USA) using a 4-pt fixture made of Al_2O_3. Before applying the load during testing at high temperature, a dwell time of 18 min was set to reach thermal equilibrium. For each temperature, at least 3 bars were tested.

The surface topological characterization of the samples used for optical measurement was carried out with a non-contact 3D profilometer (Taylor-Hobson CCI MP, Leicester, UK), equipped with a 20X magnification objective lens. Two distinct areas of 0.08×1 cm^2, were scanned at the center of the samples along two orthogonal directions and the topography data were analysed with the software Talymap 6.2 (Taylor-Hobson, Leicester, UK). The evaluation of 2D texture parameters was performed on 6 different profiles (3 for each area) extracted from the 3D data. The cut-off (λc, gaussian filter) applied for the separation of the roughness and waviness components was set according to the ISO 4288:2000 [34]. The 2D parameters were calculated as an average of the estimated values on all sampling lengths over each profile.

The 3D parameters were evaluated on the two scanned areas after removal of the surface form, short-wave and long-wave surface components (S-L surfaces) by applying a fifth order polynomial, 2.5 μm S-filter and 0.08 mm L-filter nesting indexes, respectively.

Hemispherical reflectance spectra were acquired using a double-beam spectrophotometer (Lambda900 by Perkin Elmer, Waltham, MA, USA) equipped with a Spectralon®-coated integration sphere (Perkin Elmer, Waltham, MA, USA) for the 0.25–2.5 μm wavelength region and a Fourier Transform spectrophotometer (FT-IR "Excalibur" by Bio-Rad, Hercules, CA, USA) equipped with a gold-coated integrating sphere and a liquid nitrogen-cooled detector for the range 2.5–16.5 μm.

3. Results and Discussion

3.1. Powder Synthesis and Consolidation

Neither compositional nor crystallite size changes were observed after mixing the elemental powders before the SHS stage. This feature is strictly ascribed to the extremely mild milling conditions correspondingly adopted, *i.e.*, CR = 0.14, and t_M = 20 min. For the sake of comparison, it should be noted that when Zr and graphite powders were mechanically treated using a CR value of about 5.7 [5], line reflections of carbon disappeared from the XRD patterns after 2 h of milling, and the occurrence of an exothermic reaction was evident 20 min later. Such conditions are much more severe with respect to those applied in the present work, where milling of reactants was only aimed at obtaining a homogeneous mixture to be processed by SHS.

The favorable formation enthalpy of reaction (1), *i.e.*, ($\Delta H_f^0 = -196.648$ kJ/mol) [35], makes this reacting system prone to evolve under the combustion regime. In this regard, it should be noted that the synthesis and the simultaneous consolidation of ZrC from its elements might be not be convenient, as recently pointed out for similarly behaving systems processed by reactive SPS [28]. Indeed, when the reaction for the synthesis of ZrB_2 was allowed to occur under the combustion regime, the impurities initially present in the raw powders gave rise to the sudden formation of gases, which could only barely escape from the confined graphite crucible. As a result, non-homogeneous microstructures and residual porosities were present in the obtained bulk products. Moreover, additional drawbacks (die/plungers breakage, safety problems, *etc.*) arose during the process. Based on such considerations, the synthesis by SHS of zirconium carbide and the densification of the resulting powders were carried out in two distinct steps in this work.

The high exothermic character of reaction (1) was confirmed by the self-propagating behavior exhibited by the reaction system as the pellet was locally ignited. Specifically, the generated combustion front spontaneously travelled through the specimen with average velocity and combustion temperature values equal to 8.5 ± 0.1 mm/s and 2390 ± 80 °C, respectively. The latter values are relatively lower with respect to those ones reported in previous studies related to the synthesis of ZrC by SHS from elemental reactants [9–11]. For instance, front velocity and combustion temperature of 1 cm/s and 2900 °C, respectively, were obtained by Lee *et al.* [9] in presence of 15 wt % ZrC as diluent. Moreover, combustion temperature values of up to 3100 °C were recorded by Song *et al.* [10] during the synthesis process. Such discrepancies can be mainly ascribed to the use of carbon black instead of graphite as starting reactant. Indeed, relatively higher reactivity, with respect to the synthesis carried out using graphite, is clearly obtained in the first case because of the amorphous nature and the finer particle sizes (down to the nano-scale) of C black. In addition, the high-energy ball milling treatment received by the initial mixture in Song *et al.* [10] investigation further promoted powder reactivity.

The XRD analysis of the synthesis product shown in Figure 1 indicates that, during SHS, the original reactants are completely converted with the exclusive formation of the desired cubic ZrC phase. It should be also noted that, in accordance with the typical self-cleaning character displayed by SHS processes [36], the small amount of unidentified impurities initially present in the starting powders were not detected in XRD spectra of the synthesis product.

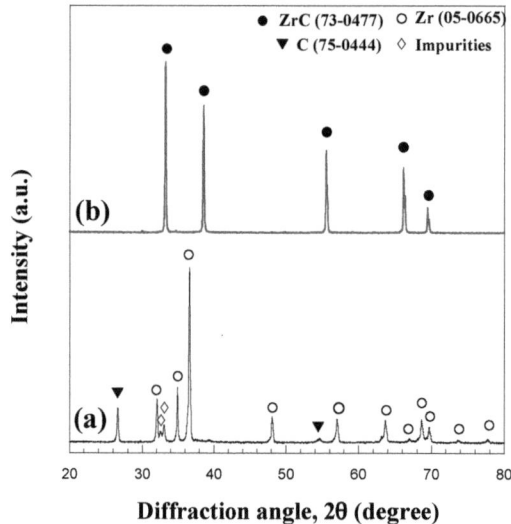

Figure 1. XRD patterns of starting reactants (**a**); and SHS product (**b**).

The powders obtained after ball milling the SHS product for different time intervals have been characterized in terms of particles size and morphology. The results related to the granulometry measured by laser light scattering analysis are summarized in Table 1. As expected, all of the particles size parameters decreased as the milling time was increased from 5 to 20 min. In particular, the average diameter (d_{av}) was correspondingly reduced from about 12.5 to less than 5 μm. In contrast, a peculiar behavior was observed when the t_M value was further prolonged to 2 h. Indeed, while the d_{10} and d_{50} values still decreased, to indicate that the smaller particles were monotonically reduced during milling, the d_{90} parameter increased with respect to that measured after 20 min milling time. Correspondingly, the resulting average diameter value (about 9 μm) was also relatively higher as compared to the measurement obtained at the previous time point.

Table 1. Particle size characteristics, as determined by laser scattering analysis, and crystallite size, obtained using the Rietveld analytical procedure, of differently ball-milled SHSed powders.

t_M (min)	d_{10} (μm)	d_{50} (μm)	d_{90} (μm)	d_{av} (μm)	Crystallite Size (nm)
5	1.50 ± 0.06	9.20 ± 0.11	29.90 ± 1.65	12.60 ± 0.25	>200
20	0.61 ± 0.03	3.68 ± 0.01	10.12 ± 0.21	4.59 ± 0.04	137.3 ± 1.2
120	0.11 ± 0.01	1.55 ± 0.32	34.10 ± 5.85	8.99 ± 1.58	44.2 ± 0.1

An increase in particle size, as revealed by laser-scattering analysis, was also recently observed during ball milling of commercial ZrC powders [20]. In particular, particle agglomeration was displayed when, after sufficiently long time milling treatments, the crystallite size was reduced at the nanoscale. Such outcome was attributed to cold-welding phenomena provoked by impacts taking place between milling media and powders. The occurrence of cold-welding is typically observed during milling of metal powders, as a consequence of their plastic deformation. On the other hand, it is less likely that such a mechanism has a major role when hard refractory ceramics like ZrC, which exhibit a fragile behavior, are mechanically processed, at least under the experimental conditions adopted in the present work.

To better clarify this issue, the milled powders were also examined by SEM, as reported later.

Iron contamination of milled powders from vials and balls, both made of steel, was also determined. The obtained values were less than 0.06 wt % and 0.12 wt % for the cases of ZrC powders milled for 20 min and 120 min, respectively. These values are both rather low, albeit iron contamination tends to become significant when the mechanical treatment is carried out for 2 h or longer times.

The SEM images reported in Figure 2a indicate that powders processed for 5 min mainly consist of large particles whose size is even larger than 10 μm, while a minor amount of few micrometer sized grains is detected. Particle refinement below 10 μm clearly took place when the milling time was increased from 5 to 20 min (cfr. Figure 2b). These outcomes are perfectly consistent with the results obtained by laser scattering analysis, as reported in Table 1. The SEM micrograph shown in Figure 2c indicates that when the milling treatment was prolonged to 2 h, the relative amount of smaller particles, whose size was further reduced, increased. In addition, this feature is in agreement with data reported in Table 1. On the other hand, some discrepancies are found when examining the relatively larger particles obtained for t_M = 2h. Specifically, the maximum size of each individual particle, generally less than 10 μm, changed only modestly with respect to those ones resulting after 20 min milling. However, the difference is that, in the first case (t_M = 2 h), particles are extensively covered by several submicron-sized grains. This feature explains why laser scattering analysis provided higher d_{90} and d_{av} values (cfr. Table 1) with respect to 20 min milled powders, despite of the prolonged ultrasonic treatment carried out. Based on these observations, the formation of such aggregates is more probably due to electrostatic charging phenomena induced by the milling process, although the contribution of cold-welding phenomenon is not completely excluded.

Figure 2. SEM images of SHS ZrC powders milled at different time intervals (t_M): 5 min (**a**); 20 min (**b**); and 120 min (**c**).

Table 1 also shows the progressive crystallite refinement of ZrC powders during the milling treatment. Specifically, it is seen that the powders mechanically processed for 20 min and 2 h display average crystallite sizes of about 140 nm and 45 nm, respectively. On the other hand, the size of 252 nm resulting from the application of the Rietveld analytical procedure on the XRD pattern of 5 min milled powders was above the threshold limit (about 200 nm) for this method. The effect of milling treatment of commercial ZrC powders was also recently taken into account by Núñez-Gonzalez *et al.* [20]. The relatively smaller crystallite size, down to 20 nm, obtained in the latter study for similar milling time conditions can be ascribed not only to the differences in the starting powders (3 μm particles sized), but, mainly, to the significantly more intense mechanical treatment, as demonstrated by the charge ratio equal to four, which is twice with respect to the value adopted in the present work.

The more convenient milling time (t_M), in the range 5–120 min, for the SHS powders to be consolidated was then identified under prescribed SPS conditions, *i.e.*, T_D = 1850 °C and t_D = 20 min. The obtained results shown in Figure 3 indicate that an increase of the t_M value from 5 to 20 min gives rise to a marked enhancement in powder densification. In contrast, only modest benefits are obtained when the powders milled for 2 h were consolidated by SPS. This outcome can be associated to the corresponding particle size evolution during the mechanical treatment (*cfr.* Figure 2 and Table 1), *i.e.*, the main changes are mostly confined to the milling time interval 5–20 min.

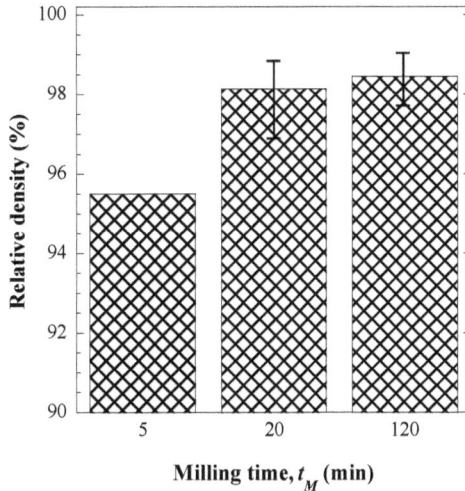

Figure 3. Effect of milling time on the relative density of ZrC samples obtained by SPS ($T_D = 1850\ °C$, $t_D = 10$ min, $p = 50$ MPa).

Along this line, it was recently reported that the SPS conditions needed to obtain nearly fully dense zirconium diboride and carbide products were significantly affected by the mechanical treatment previously received by the powders, only if crystallite size is reduced at the nanoscale [20,37]. In particular, the temperature level required to achieve, under 75 MPa mechanical pressure, the complete or near-complete densification of ZrC was lowered from 2100 °C (3 min holding time), for the case of as-received powders, to about 1850 °C (no socking time), if the crystallite size was decreased down to 20 nm through a high-energy ball milling treatment carried out up to 3 h with CR = 4 [20]. Such results are quite consistent to those ones obtained in the present work, where similar densification levels are also reached at 1850 °C when using 20 min or 120 min milled powders. The shorter sintering time adopted by Núñez-Gonzalez *et al.* [20] could be likely explained by the relatively finer crystallite size of the powders processed by SPS, as a consequence of the more severe milling treatment applied. It should be noted that higher milling intensities were not considered in the present work to avoid an excessive iron contamination from milling media, which would negatively affect the high-temperature properties of the resulting material. According to the consideration above, only the SHS powders milled for 20 min have been processed by SPS hereinafter, and the related results will be described and discussed in what follows.

Figure 4a shows the effect of the sintering temperature in the 1750–1850 °C range on the consolidation level when maintaining constant the holding time ($t_D = 20$ min), and applied pressure (50 MPa). As expected, higher dense samples are produced as the sintering temperature was progressively augmented. Along the same direction, the data plotted in Figure 4b indicate that an increase of the holding time from 5 to 20 min significantly promotes the consolidation of ZrC powders. Nonetheless, no additional beneficial effects are obtained when the processing time was further prolonged.

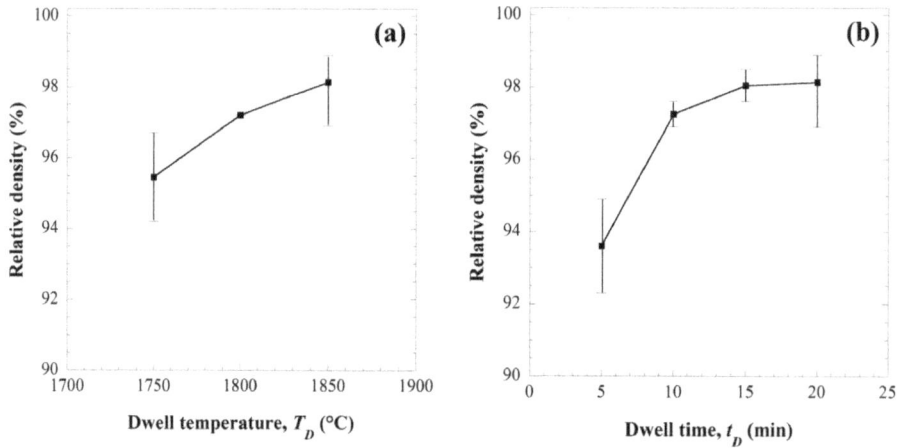

Figure 4. Effect of (**a**) the sintering temperature (t_D = 20 min, p = 50 MPa, t_H = 10 min) and (**b**) dwell time (T_D = 1850 °C, p = 50 MPa, t_H = 10 min) on the density of ZrC samples obtained by SPS.

Thus, within the experimental conditions considered in the present work, the optimal T_D and t_D values able to guarantee the fabrication of about 98% (average value) dense ZrC products by SPS are 1850 °C and 20 min, respectively. It should be noted that such conditions are among the mildest one reported in the literature for the obtainment of similar density levels by SPS when using commercial ZrC powders [8,18–20,23]. The only exception is represented by the slightly lower temperatures (1800 °C) recently adopted by Wei *et al.* [23] to obtain 97.87% dense samples when taking advantage of a special double-die tooling setup, which enables the application of considerably higher mechanical pressures (200 MPa).

A SEM micrograph showing the fracture surface of a dense ZrC sample obtained by SPS at T_D = 1850 °C and t_D = 20 min is reported in Figure 5. This image shows a transgranular fracture and a good level of consolidation, although a residual amount of closed porosity is present, with pore size in the range 1–3 μm.

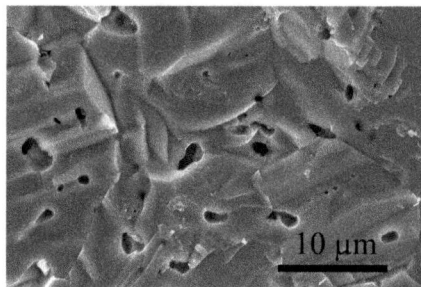

Figure 5. SEM image of the fracture surface of ZrC product obtained by SPS (T_D = 1850 °C, p = 50 MPa, t_H = 10 min, t_D = 20 min).

A backscattered image collected on the polished surface (Figure 6a) shows the presence of black spots. From the In-lens image (Figure 6b), it can be appreciated that the latter ones are either pores or residual carbon inclusions (see EDS spectrum). The total amount of carbon inclusion and pores is, however, below 7 vol %.The mean grain size of ZrC grains is around 15 ± 5 μm, with minimum grains size around 8 μm and maximum value around 28 μm, *i.e.*, higher than monolithic ZrC obtained through SPS of commercial ZrC powders, which resulted to be about 13 μm [18]. It is presumed that the ultrafine powder fraction coalesced on the coarser particles to generate larger final grains when exposed to the high temperature conditions established during the sintering process.

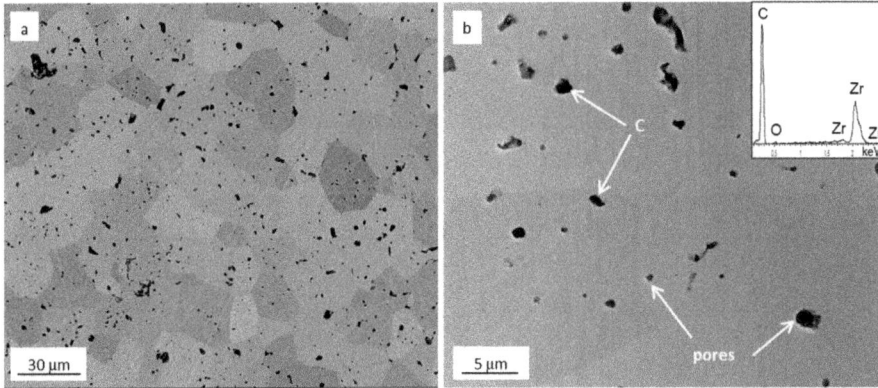

Figure 6. SEM images of the polished surface of the SHS-SPSed ZrC sample showing (**a**) a microstructure overview in back-scattered image mode and (**b**) the presence of closed porosity and carbon inclusions in In-lens mode with the EDS spectrum of C-rich pockets inset.

As reported in the Section 2, the 40 mm diameter samples needed for the evaluation of thermo-mechanical properties, as reported in the next section, have been produced taking advantage of an SPS apparatus (FCT Systeme GmbH, Rauenstein, Germany), different from that one (Sumitomo, Kanagawa, Japan) considered so far to obtain 14.7 mm sized specimens, the latter one being unable to provide the required higher electric current levels. In this regard, it should be noted that, although the two pieces of equipment are based on the same principle, they display some differences in the temperature measurements. In particular, for the Sumitomo machine, the temperature is monitored and controlled at the surface of the die, while such measurement is made at an axial point for the FCT equipment. Thus, to compensate the thermal gradients present along the radial direction, the dwell temperature value set in the latter case was increased from 1850 °C to 1950–2000 °C, while the holding time and the applied pressure remained unchanged, *i.e.*, 50 MPa and 20 min, respectively. Correspondingly, the obtained products are characterized by density of 97.3% or higher values, which fall within the error bar of the data reported in Figure 4 for smaller diameter samples. Moreover, SEM analysis performed on the resulting SPSed products indicated that the ZrC grains size were also comparable in the two cases. Based on these findings, it is possible to state that SHS powders undergoing sintering are subjected to similar thermal cycles in the two machines.

3.2. Mechanical Characterization

The hardness of monolithic ZrC samples obtained in this work by SPS was 17.5 ± 0.4 GPa. This value is in very good agreement with those previously found on monophasic ZrC produced through SPS from commercial powders, *i.e.*, 17.9 ± 0.6 GPa [18]. Higher values could be likely obtained if the mean grains size, residual porosity and carbon inclusions are significantly reduced.

As for fracture toughness, the resulting mean K_{IC} value, 2.6 ± 0.2 MPa·$m^{0.5}$, is slightly larger than the one found for ZrC-based materials containing 1 vol % $MoSi_2$, *i.e.*, 2.1 ± 0.2 MPa·$m^{0.5}$ [18]. The brittleness of this material is also testified by the tendency of chipping of the surface during loading or immediately after the releasing of the load.

The room temperature strength of the ZrC products resulting from SHSed powders was 248 ± 66 MPa, in the same order of the values reported in the literature for ZrC-based ceramics produced by pressureless sintering [13]. Nonetheless, it should be noted that these values are significantly lower than those found for monolithic ZrC obtained by SPS from commercial powder (H.C. Starck, Grade B, Karlsruhe, Germany) [18], 407 ± 38 MPa, although, in the latter case, strength data refer to three-point bending strength performed on smaller bars (1.0 mm $\times$ 0.8 mm $\times$ 10 mm). Such discrepancies between monolithic samples obtained from SHS or commercial powders could be due to the different grain size distribution of the two materials, as mentioned above, and to the presence of different volume amount of free carbon, 5 vol % and 2 vol %, respectively. However, it should be noted, that the bending strength of the ceramic produced by SHS/SPS remained basically unaltered, *i.e.*, 247 ± 36 MPa, when the test was carried out at 1000 °C. Only when the testing temperature was further increased to 1200 °C was a notable drop down to 85 MPa observed. Correspondingly, the monolithic sample resulted as being strongly exfoliated at such a temperature, probably owing to residual traces of oxygen remained trapped in the furnace during the test.

3.3. Optical Properties

Before performing the optical measurements, the polished sample surface was preliminarily characterized from the topological point of view. The mean values of the 2D and 3D surface texture parameters, obtained according to the ISO 4287:1997 [38] and ISO 25178-2:2012 [39], respectively, are reported in Table 2 along with the corresponding standard deviations.

Table 2. Description and average (Av.) values of the measured two-dimensional (2D) and three-dimensional (3D) surface parameters.

2D Parameters	Av. Values	Description
Ra (μm)	0.009 ± 0.004	Arithmetic Mean Deviation of the roughness profile.
Rq (μm)	0.014 ± 0.006	Root-Mean-Square (RMS) Deviation of the roughness profile.
Rsk	-0.91 ± 0.15	Skewness of the roughness profile.
Rku	4.90 ± 0.56	Kurtosis of the roughness profile.
Rp (μm)	0.020 ± 0.012	Maximum Peak Height of the roughness profile.
Rv (μm)	0.047 ± 0.017	Maximum Valley Depth of the roughness profile.
Rz (μm)	0.068 ± 0.028	Maximum height of roughness profile on the sampling length.
Rc (μm)	0.043 ± 0.020	Mean height of the roughness profile elements.
Rt (μm)	1.29 ± 0.77	Total height of roughness profile (on the evaluation length).
RSm (mm)	0.076 ± 0.012	Mean Width of the roughness profile elements.
Rdq	0.43 ± 0.19	Root-Mean-Square Slope of the roughness profile.
3D Parameters	**Av. Values**	**Description**
Sa(μm)	0.070 ± 0.030	Arithmetic mean height of the S-L Surface.
Sq(μm)	0.56 ± 0.16	Root mean square height of the S-L Surface.
Ssk	-11.4 ± 5.2	Skewness of the S-L Surface.
Sku	150 ± 70	Kurtosis of the S-L Surface.
Sp(μm)	11.0 ± 3.10	Maximum peak height in the S-L Surface.
Sv(μm)	19.6 ± 1.60	Maximum pit height of the S-L Surface.
Sz(μm)	30.6 ± 4.7	Maximum height of the S-L Surface.

Ra and Rq represent the most commonly amplitude parameters used in surface texture characterization and provide the average roughness. The Rq parameter has a more statistical significance with respect to Ra and is related, as the 3D height parameter Sq, to the surface energy and to the amount of light scattered from smooth surfaces [40]. The analyzed surfaces show very low values of both Ra (0.009 μm) and Rq (0.014 μm).

However, Ra and Rq do not give information about the shape of surface irregularities and the abnormal peaks or valleys do not affect significantly their values. The skewness and kurtosis parameters are related to defects distribution of the studied surface: negative values of Rsk indicate the prevalence of pores or valleys, while Rku values higher than three indicate the presence of sharp defects. Specifically, the values of these parameters for the ZrC samples show that the distribution of the extracted profile heights is asymmetric, negative (Rsk = −0.91) and quite narrow, (Rku = 4.9). The parameter Rdq is related to the optical properties of the surface and low values, as for the present case (Rdq = 0.43), specify that the surface is a good reflector. The profile slope is very low. Therefore, the measured 2D amplitude parameters indicate that the studied surface is very smooth but is characterized by the presence of valleys or holes.

The values of the 3D parameters reported in Table 2 are higher than the corresponding 2D, probably because of the larger sampling area that may include a high number of surface defects. Particularly high is the value of the Sku parameter. This outcome may be due to the fact that this index is extremely sensitive to local defects (holes or valleys) and also to error propagation, since it depends on high order powers in its mathematical expression. In any case, the 3D parameters are in substantial agreement with the corresponding 2D parameters and confirm the previous findings.

Figure 7 shows the hemispherical reflectance of the ZrC specimen. It is possible to appreciate the typical step-like spectrum of UHTC carbides, characterized by a low reflectance (*i.e.*, a high optical absorption) in the visible-near infrared and a high reflectance plateau (*i.e.*, a low thermal emittance) at longer infrared wavelengths.

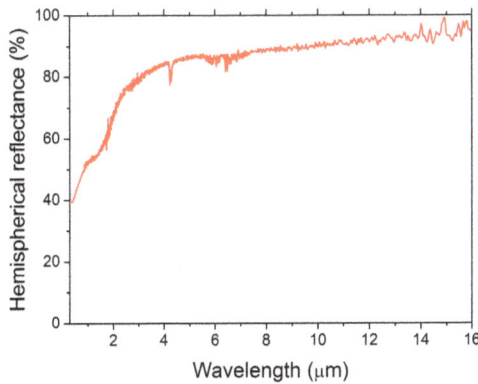

Figure 7. Hemispherical reflectance spectrum of ZrC sample.

For solar absorber applications, sunlight absorption characteristics are quantified by the total directional solar absorbance at the temperature T:

$$\alpha\prime_S(T) = \frac{\int\limits_{\lambda_{min}}^{\lambda_{max}} (1 - \rho\prime^\cap(\lambda, T)) \times S(\lambda)d\lambda}{\int\limits_{\lambda_{min}}^{\lambda_{max}} S(\lambda)d\lambda} \tag{2}$$

which is defined in terms of the spectral directional-hemispherical reflectance $\rho\prime^\cap(\lambda,T)$ (*i.e.*, the hemispherical reflectance acquired for a given incidence direction of light, like it happens for measurements carried out with an integrating sphere) and of sunlight spectrum $S(\lambda)$ [41]. The energy lost by thermal radiation by the absorber heated at the temperature T is connected to the total directional thermal emittance parameter expressed by:

$$\varepsilon\prime_{\lambda1,\lambda2}(T) = \frac{\int\limits_{\lambda1}^{\lambda2} (1 - \rho\prime^\cap(\lambda, T)) \times B(\lambda, T)d\lambda}{\int\limits_{\lambda1}^{\lambda2} B(\lambda, T)d\lambda} \tag{3}$$

where $B(\lambda,T)$ is the blackbody spectrum at the temperature T. The α/ε ratio is called spectral selectivity and is connected to the material ability to respond differently to optical radiation in the sunlight and thermal emission spectral ranges. Thus, roughly speaking, an ideal solar absorber should have a high $\alpha \approx 1$, a low $\varepsilon \approx 0$ and the highest as possible α/ε parameter. Under the remarks made by the authors in a recent work [42], the $\rho'^{\cap}(\lambda,T)$ function can be approximated with the spectral hemispherical reflectance measured at room temperature.

For the sake of comparison of the sample tackled in the present work with analogous previously investigated systems, *i.e.*, sintering aid-doped ZrC [43] and fully dense monolithic ZrB$_2$ obtained by SHS/SPS [44]. Parameters appearing in Equations (2) and (3) have been calculated by considering integration bounds $\lambda_{min} = 0.3$ μm, $\lambda_{max} = 2.3$ μm, $\lambda_1 = 0.3$ μm, $\lambda_2 = 15.0$ μm and temperatures of 1200 K and 1400 K. Correspondingly, $\alpha = 0.51$, $\varepsilon(1200\ K) = 0.21$, and $\varepsilon(1400\ K) = 0.24$ are obtained for the case of the ZrC sample produced in this work, which leads to $\alpha/\varepsilon(1200\ K) = 2.4$, and $\alpha/\varepsilon(1400\ K) = 2.1$. Based on these results, it can be stated that the latter product is slightly less absorptive and emissive with respect to MoSi$_2$-containing ZrC samples ($\alpha = 0.55$–0.56 and $\varepsilon(1200\ K) = 0.23$–0.27), while spectral selectivity (2.4–2.1) is similar [43]. In addition, when the comparison is extended to the bulk additive-free ZrB$_2$ material recently processed following the same SHS/SPS route [40], it is found that the ZrC product appears slightly more absorptive and emissive, *i.e.*, $\alpha_{ZrB2} = 0.47$, $\varepsilon_{ZrB2}(1400\ K) = 0.18$, with a slightly higher spectral selectivity displayed by the diboride system ($\alpha/\varepsilon = 2.6$).

Finally, it should be mentioned that a possible criticism for UHTCs in solar applications is represented by the relatively low solar absorbance when comparing them, for instance, to silicon carbide. However, it has been recently demonstrated,in the case of hafnium carbide [45], that solar absorbance can be significantly increased by surface texturing without detrimentally affecting thermal emittance.

4. Conclusions

ZrC-based ultra-refractory ceramics are promising candidate materials for solar energy and application in other industrial fields. In spite of their technological interest, the conventional processing technologies generally utilized to obtain highly dense products require severe sintering conditions (temperatures above 2000 °C and dwelling times on the order of hours), unless sintering aids are introduced. In this work, about 98% dense monolithic ZrC samples are successfully produced by SPS after 20 min holding time at 1850 °C, being the starting powders preliminarily synthesized by SHS followed by 20 min ball milling treatment. The milder sintering conditions adopted in this work can be taken as an indication of the high sintering ability of the powder produced by SHS.

The measured Vickers hardness and fracture toughness of the resulting material, 17.5 ± 0.4 GPa and 2.6 ± 0.2 MPa· m$^{0.5}$, respectively, are similar to the values reported in the literature for ZrC monoliths. The relatively low mechanical strength, about 250 MPa at room temperature, is likely associated to the coarse microstructure characterizing the bulk ceramic, with grain size up to 28 μm, as well as to the presence of residual pores and carbon inclusions. Nevertheless, the obtained room temperature strength remains basically unchanged up to 1000 °C, whereas such a property becomes significantly worse (85 MPa) as the temperature is further increased to 1200 °C. In this regard, the use of higher purity and finer initial powders are expected to improve the densification as well as the microstructure of the sintered product and, in turn, the resulting strength.

The topological 2D and 3D characterization of the polished sample exposed to optical measurements indicate that the studied surface is very smooth, albeit valleys or holes are also present.

Regarding the optical behavior, the obtained ZrC specimen displayed a step-like spectrum, typical of transition metal carbides, with high absorption in the visible-near infrared area and low thermal emittance at longer infrared wavelengths. Furthermore, the resulting spectral selectivity values are sufficiently good, 2.4 and 2.1 at 1000 K and 1400 K, respectively.

In summary, it is possible to conclude that the combination of the SHS and SPS routes provide highly dense ZrC products with performances for solar absorber applications comparable with those displayed by analogous UHTC materials, specifically monolithic ZrB_2 obtained with the same synthesis/sintering techniques and additive-containing ZrC produced by alternative methods.

Acknowledgments: The financial support from the Italian Ministry of Education, Universities and Research, MIUR (Italy) through the project (Prot. RBFR12TIT1) FIRB 2012 "Futuro in Ricerca" is gratefully acknowledged. IM (Innovative Materials) S.r.l., Italy, is also acknowledged for granting the use of HPD 25-1 apparatus (FCT Systeme GmbH, Rauenstein, Germany) utilized for the preparation of 40 mm diameter specimens. One of us (Elisa Sani) acknowledges the Italian bank foundation "Fondazione Ente Cassa di Risparmio di Firenze" for providing the grant for Marco Meucci within the framework of the "SOLE" and "SOLE-2" projects (pratiche n. 2013.0726 and 2014.0711). Thanks are due to Massimo D'Uva and Mauro Pucci (CNR-INO) for technical assistance.

Author Contributions: Clara Musa, Roberta Licheri, Roberto Orrù and Giacomo Cao conceived, designed and performed SHS and SPS experiments and analyzed the related data; Diletta Sciti, Laura Silvestroni and Luca Zoli carried out microstructural and thermomechanical characterization of the obtained products and analyzed the corresponding data; Andrea Balbo performed surface texture measurements and analyzed the related data; optical properties and the analysis of the obtained results was performed by Luca Mercatelli, Marco Meucci and Elisa Sani. All the authors discussed the results and wrote the paper.

Conflicts of Interest: The authors declare no conflict of interest. The founding sponsors had no role in the design of the study; in the collection, analyses, or interpretation of data; in the writing of the manuscript, and in the decision to publish the results.

References

1. Upadhya, K.; Yang, J.-M.; Hoffman, W.P. Materials for ultrahigh temperature structural applications. *Am. Ceram. Soc. Bull.* **1997**, *76*, 51–56.

2. Katoh, Y.G.; Vasudevamurthy, T.; Nozawa, L.L. Snead Properties of zirconium carbide for nuclear fuel applications. *J. Nucl. Mater.* **2013**, *441*, 718–742. [CrossRef]

3. Sani, E.; Mercatelli, L.; Francini, F.; Sans, J.-L.; Sciti, D. Ultra-refractory ceramics for high-temperature solar absorbers. *Scr. Mater.* **2011**, *65*, 775–778. [CrossRef]

4. Maitre, A.; Lefort, P. Solid state reaction of zirconia with carbon. *Solid State Ion.* **1997**, *104*, 109–122. [CrossRef]

5. Yen, B.K. X-ray diffraction study of mechanochemical synthesis and formation mechanisms of zirconium carbide and zirconium silicides. *J. Alloys Compd.* **1998**, *268*, 266–269. [CrossRef]

6. Shen, G.; Chen, D.; Liu, Y.; Tang, K.; Qian, Y. Synthesis of ZrC hollow nanospheres at low temperature. *J. Cryst. Growth* **2004**, *262*, 277–280. [CrossRef]

7. Li, J.; Fu, Z.Y.; Wang, W.M.; Wang, H.; Lee, S.H.; Niihara, K. Preparation of ZrC by self-propagating high-temperature synthesis. *Ceram. Int.* **2010**, *36*, 1681–1686. [CrossRef]

8. Xie, J.; Fu, Z.; Wang, Y.; Lee, S.W.; Niihara, K. Synthesis of nanosized zirconium carbide powders by a combinational method of sol-gel and pulse current heating. *J. Eur. Ceram. Soc.* **2014**, *34*. [CrossRef]

9. Lee, H.B.; Cho, K.; Lee, J.W. Synthesis and temperature profile analysis of ZrC by SHS method. *J. Korean Ceram. Soc.* **1995**, *32*, 659–668.

10. Song, M.S.; Huang, B.; Zhang, M.X.; Li, J.G. In situ synthesis of ZrC particles and its formation mechanism by self-propagating reaction from Al–Zr–C elemental powders. *Powder Technol.* **2009**, *191*, 34–38. [CrossRef]

11. Zhang, M.X.; Hu, Q.D.; Huang, B.; Li, J.G. Fabrication of ZrC particles and its formation mechanism by self-propagating high-temperature synthesis from Fe–Zr–C elemental powders. *J. Alloys Compd.* **2011**, *509*, 8120–8125. [CrossRef]

12. Song, M.; Ran, M.; Long, Y. Synthesis of ultrafine zirconium carbide particles by SHS in an Al–Zr–C system: Microstructural evaluation and formation mode. *J. Alloys Compd.* **2013**, *564*, 20–26. [CrossRef]

13. Silvestroni, L.; Sciti, D. Microstructure and properties of pressureless sintered ZrC-based materials. *J. Mater. Res.* **2008**, *23*, 1882–1889. [CrossRef]

14. Zhao, L.; Jia, D.; Duan, X.; Yang, Z.; Zhou, Y. Pressureless sintering of ZrC-based ceramics by enhancing powder sinterability. *Int. J. Refract. Met. Hard Mater.* **2011**, *29*, 516–521. [CrossRef]

15. Nachiappan, C.; Rangaraj, L.; Divakar, C.; Jayaram, V. Synthesis and Densification of monolithic zirconium carbide by reactive hot pressing. *J. Am. Ceram. Soc.* **2010**, *93*, 1341–1346. [CrossRef]

16. Wang, X.-G.; Liu, J.-X.; Kan, Y.-M.; Zhang, G.-J. Effect of solid solution formation on densification of hot-pressed ZrC ceramics with MC (M = V, Nb and Ta) additions. *J. Eur. Ceram. Soc.* **2012**, *32*, 1795–1802. [CrossRef]

17. Orrù, R.; Licheri, R.; Locci, A.M.; Cincotti, A.; Cao, G. Consolidation/synthesis of materials by electric current activated/assisted sintering. *Mater. Sci. Eng. R* **2009**, *63*, 127–287. [CrossRef]

18. Sciti, D.; Guicciardi, S.; Nygren, M. Spark Plasma Sintering and mechanical behavior of ZrC-based composites. *Scr. Mater.* **2008**, *59*, 638–641. [CrossRef]

19. Gendre, M.; Maitre, A.; Trolliard, G. A study of the densification mechanism during spark plasma sintering of zirconium (oxy-) carbide powders. *Acta Mater.* **2010**, *58*, 2598–2609. [CrossRef]

20. Núñez-Gonzalez, B.; Ortiz, A.L.; Guiberteau, F.; Nygren, M. Improvement of the Spark-Plasma-Sintering Kinetics of ZrC by High-Energy Ball-Milling. *J. Am. Ceram. Soc.* **2012**, *95*, 453–456. [CrossRef]

21. Sun, S.-K.; Zhang, G.-J.; Wu, W.-W.; Liu, J.-X.; Suzuki, T.; Sakka, Y. Reactive spark plasma sintering of ZrC and HfC ceramics with fine microstructures. *Scr. Mater.* **2013**, *69*, 139–142. [CrossRef]

22. Bertagnoli, D.; Borrero-López, O.; Rodríguez-Rojas, F.; Guiberteau, F.; Ortiz, A.L. Effect of processing conditions on the sliding-wear resistance of ZrC triboceramics fabricated by spark-plasma sintering. *Ceram. Int.* **2015**, *41*, 15278–15282. [CrossRef]

23. Wei, X.; Back, C.; Izhanov, O.; Khasanov, O.L.; Haines, C.D.; Olevsky, E. Spark Plasma Sintering of Commercial Zirconium Carbide Powders: Densification Behaviour and Mechanical Properties. *Materials* **2015**, *8*, 6043–6061. [CrossRef]

24. Mishra, S.K.; Das, S.; Pathak, L.C. Defect structures in zirconium diboride powder prepared by self-propagating high-temperature synthesis. *Mater. Sci. Eng. A* **2004**, *364*, 249–255. [CrossRef]

25. Licheri, R.; Orrù, R.; Musa, C.; Cao, G. Combination of SHS and SPS Techniques for Fabrication of Fully Dense ZrB$_2$-ZrC-SiC Composites. *Mater. Lett.* **2008**, *62*, 432–435. [CrossRef]

26. Cincotti, A.; Licheri, R.; Locci, A.M.; Orrù, R.; Cao, G. A review on combustion synthesis of novel materials: Recent experimental and modeling results. *J. Chem. Technol. Biotechnol.* **2003**, *78*, 122–127. [CrossRef]

27. Musa, C.; Orrù, R.; Licheri, R.; Cao, G. Spark Plasma Synthesis and Densification of TaB$_2$ by Pulsed Electric Current Sintering. *Mater. Lett.* **2011**, *65*, 3080–3082. [CrossRef]

28. Licheri, R.; Musa, C.; Orrù, R.; Cao, G. Influence of the heating rate on the in-situ synthesis and consolidation of ZrB$_2$ by Reactive Spark Plasma Sintering. *J. Eur. Ceram. Soc.* **2015**, *35*, 1129–1137. [CrossRef]

29. Haynes, W.M. *CRC Handbook of Chemistry and Physics*, 93rd ed.; CRC Press: Boca Raton, FL, USA, 2012.

30. Lutterotti, L.; Ceccato, R.; Dal Maschio, R.; Pagani, E. Quantitative analysis of silicate glass in ceramic materials by the Rietveld method. *Mater. Sci. Forum* **1998**, *278*, 87–92. [CrossRef]

31. Evans, A.G.; Charles, E.A. Fracture toughness determination by indentation. *J. Am. Ceram. Soc.* **1976**, *59*, 371–372. [CrossRef]

32. *Advanced Technical Ceramics—Monolithic Ceramics—Mechanical Properties at Room Temperature—Part 1: Determination of Flexural Strength*; ENV843-1:2004; BSI: London, UK; November 2004.

33. *Advanced Technical Ceramics—Methods for Testing Monolithic Ceramics—Thermomechanical Properties—Part 1: Determination of Flexural Strength at Elevated Temperatures*; EN 820-1:2002; BSI: London, UK; March 2003.

34. *Geometrical Product Specifications (GPS)—Surface Texture: Profile Method—Rules and Procedures for the Assessment of Surface Texture*; ISO 4288:2000/Cor.1:1998; ISO: Geneva, Switzerland; June 1998.

35. Barin, I. *Thermochemical Data of Pure Substances*; VHC: Weinheim, Germany, 1989.

36. Varma, A.; Rogachev, A.S.; Mukasyan, A.S.; Hwang, S. Combustion Synthesis of Advanced Materials: Principles and Applications. *Adv. Chem. Eng.* **1998**, *24*, 79–226.

37. Zamora, V.; Ortiz, A.L.; Guiberteau, F.; Nygren, M. Crystal-size dependence of the spark-plasma-sintering kinetics of ZrB2 ultra-high-temperature ceramics. *J. Eur. Ceram. Soc.* **2012**, *32*, 271–276. [CrossRef]

38. *Geometrical Product Specifications (GPS)—Surface Texture: Profile Method—Terms, Definitions and Surface Texture Parameters*; ISO 4287:1997/Amd 1:2009; ISO: Geneva, Switzerland; June 1998.

39. *Geometrical Product Specifications (GPS)—Surface Texture: Areal—Part 2: Terms, Definitions and Surface Texture Parameters*; ISO 25178-2:2012; ISO: Geneva, Switzerland; April 2012.

40. Leach, R.K. *Fundamentals Principles of Engineering Nanometrology*, 2nd ed.; Elsevier: Amsterdam, The Netherlands, 2010.

41. *Solar Spectral Irradiance*; Technical Report, No. 85; Commission Internationale de l'Eclairage: Vienna, Austria, 1989.

42. Sani, E.; Mercatelli, L.; Sansoni, P.; Silvestroni, L.; Sciti, D. Spectrally selective ultra-high temperature ceramic absorbers for high-temperature solar plants. *J. Renew. Sustain. Energy* **2012**, *4*, 033104. [CrossRef]
43. Sani, E.; Mercatelli, L.; Meucci, M.; Balbo, A.; Silvestroni, L.; Sciti, D. Compositional dependence of optical properties of zirconium, hafnium and tantalum carbides for solar absorber applications. *Sol. Energy* **2016**, *131*, 199–207. [CrossRef]
44. Sani, E.; Mercatelli, L.; Meucci, M.; Balbo, A.; Musa, C.; Licheri, R.; Orrù, R.; Cao, G. Optical properties of dense zirconium and tantalum diborides for solar thermal absorbers. *Renew. Energy* **2016**, *91*, 340–346. [CrossRef]
45. Sciti, D.; Silvestroni, L.; Trucchi, D.M.; Cappelli, E.; Orlando, S.; Sani, E. Femtosecond laser treatments to tailor the optical properties of hafnium carbide for solar applications. *Sol. Energy Mater. Sol. Cells* **2015**, *132*, 460–466. [CrossRef]

materials

MDPI

Article

The Effect of Lithium Doping on the Sintering and Grain Growth of SPS-Processed, Non-Stoichiometric Magnesium Aluminate Spinel

Yuval Mordekovitz [†], Lee Shelly [†], Mahdi Halabi, Sergey Kalabukhov and Shmuel Hayun *

Department of Materials Engineering, Ben-Gurion University of the Negev, P.O. Box 653, Beer-Sheva 8410501, Israel; yuvalmor@post.bgu.ac.il (Y.M.); leeshel@post.bgu.ac.il (L.S.); mahdi@post.bgu.ac.il (M.H.); kalabukh@bgu.ac.il (S.K.)

* Correspondence: hayuns@bgu.ac.il; Tel.: +972-8-642-8742; Fax: +972-8-642-8744
† These authors contributed equally to this work.

Academic Editor: Eugene A. Olevsky
Received: 11 May 2016; Accepted: 7 June 2016; Published: 16 June 2016

Abstract: The effects of lithium doping on the sintering and grain growth of non-stoichiometric nano-sized magnesium aluminate spinel were studied using a spark plasma sintering (SPS) apparatus. Li-doped nano-$MgO \cdot nAl_2O_3$ spinel ($n = 1.06$ and 1.21) powders containing 0, 0.20, 0.50 or 1.00 at. % Li were synthesized by the solution combustion method and dense specimens were processed using a SPS apparatus at 1200 °C and under an applied pressure of 150 MPa. The SPS-processed samples showed mutual dependency on the lithium concentration and the alumina-to-magnesia ratio. For example, the density and hardness values of near-stoichiometry samples ($n = 1.06$) showed an incline up to 0.51 at. % Li, while in the alumina rich samples ($n = 1.21$), these values remained constant up to 0.53 at. % Li. Studying grain growth revealed that in the Li-$MgO \cdot nAl_2O_3$ system, grain growth is limited by Zener pining. The activation energies of undoped, 0.2 and 0.53 at. % Li-$MgO \cdot 1.21Al_2O_3$ samples were 288 ± 40, 670 ± 45 and 543 ± 40 kJ·mol^{-1}, respectively.

Keywords: grain growth; lithium; magnesium aluminate spinel; precipitation; SPS

1. Introduction

Magnesium aluminate spinel ($MgO \cdot nAl_2O_3$) is an attractive ceramic material for many technological applications, owing to its combination of excellent mechanical and optical properties [1–3]. To realize and maximize its qualities, the spinel must be sintered to full density. Sintering to full density is usually a difficult goal to achieve, given the requirements of high pressure and elevated temperatures. Yet, even then, variations in powder quality and densification processes can cause optical defects [4–6]. To overcome these issues, the use of sintering additives, such as Na_3AlF_6 [7], AlF_3 [3], B_2O_3 [8], $AlCl_3$ [3], CaO [2], LiF [9–11] and $CaCO_3$ + LiF [4], has been proposed. Of these, it was established that LiF consistently allowed for the sintering of transparent spinel [6,9–12]. As such, the effect of LiF on the sintering behavior of $MgAl_2O_4$ has been extensively studied [6,9–14], including by Meir *et al.* [10] and Rosenburg *et al.* [11].

Two mechanisms were proposed to explain the enhanced sintering kinetics and improved transparency attained by the sintered parts. The first involves the formation of a liquid phase (LiF, melting point (m.p.) ~847 °C) at relatively low temperature that wets the $MgAl_2O_4$ particles and likely aids densification by particle rearrangement and liquid-phase sintering. The second mechanism was proposed to act at higher temperatures. Here, LiF decomposes and the highly reactive F$^-$ ions react with impurities (e.g., C and S), thereby cleaning/activating particle surfaces. In turn, the Li$^+$ cations react with the spinel, resulting in accelerated mass transport due to the formation of oxygen vacancies. Recently, we studied the effects of lithium on the energetics, thermal stability, and coarsening of $MgO \cdot nAl_2O_3$, as well as its solubility in two-alumina-rich spinel compositions ($n = 1.06$ and $n = 1.21$). It was established that the phase stability of Li-doped, near-stoichiometry ($n = 1.06$) spinels is size-dependent. The spinel structure was able to hold up to 1 at. % lithium at grain sizes smaller than 30 nm, whereas for larger crystallite sizes, Mg(Li,Al)O and γ-$LiAlO_2$ phases precipitated. The aluminum-rich samples ($n = 1.21$) showed greater phase stability, with decomposition occurring only above 1 at. % lithium, independent of crystallite size. The measured surface (and interface) enthalpies of $MgO \cdot 1.06Al_2O_3$, $MgO \cdot 1.21Al_2O_3$ and 0.20 at. % Li-$MgO \cdot 1.21Al_2O_3$ were 1.51 ± 0.15 (0.42 ± 0.20) Jm^{-2}, 1.17 ± 0.15 (0.32 ± 0.21) and 1.05 ± 0.12 (0.24 ± 0.18) Jm^{-2}, respectively [15]. These values are in agreement with the lower coarsening tendency of aluminum-rich spinels [15]. Spark plasma sintering is a well-established method for sintering transparent magnesium aluminate spinel [10,16–24] which combines axial pressure with heating via an electrical current passing through a die containing the powder body. A LiF sintering additive (~1 wt. %) is typically required for transparency.

In the present work, dense bodies from various lithium-doped nano-$MgO \cdot 1.06Al_2O_3$ and $MgO \cdot 1.21Al_2O_3$ spinels were SPS-processed and their microstructure and phase composition were analyzed. The cardinal role of the Li additive is emphasized and discussed.

2. Materials and Experimental Procedures

Li-doped nano-$MgO \cdot nAl_2O_3$ spinel ($n = 1.06$ and 1.21) powders containing 0, 0.20, 0.50 or 1.00 at. % Li were synthesized by the solution combustion method [25], as described in detail by Mordekovitz and Hayun [15]. A 100 mL water-based solution was prepared with the appropriate amount of magnesium nitrate ($Mg(NO_3)_2 \cdot 6H_2O$, 96% metal basis, Fluka Analytical, St. Louis, MO, USA), aluminum nitrate ($Al(NO_3)_3 \cdot 9H_2O$, 96% metal basis, Fluka Analytical) and lithium acetate ($LiCH_3CO_2 \cdot 2H_2O$, reagent grade, metal basis, Alfa Aesar, Haverhill, MA, USA). Thirty-seven grams of citric acid (ACS reagent $\geqslant 99.5$) and 6 mL ethylene glycol (anhydrous, 99.8%, Sigma Aldrich, St. Louis, MO, USA) were added to the solution. The resulting mixtures were evaporated at 120 °C under agitation by magnetic stirring until high-viscosity foam-like colloids had formed. Finally, the dried gel precursor was calcined at 850 °C for 72 h to obtain a fine powder. Sintering was conducted in a Spark Plasma Sintering Machine (FCT Systems GmbH, Rauenstein, Germany) using a modified elevated pressure set-up capable of delivering uniaxial pressures greater than 500 MPa. Ten millimeter disks were sintered using a graphite die (20 mm outer diameter) with silicon carbide (SiC) plungers placed inside a conventional 20 mm graphite die-and-plunger set. All SPS experiments were conducted in a low vacuum (1.3 hPa), with a K-type control thermocouple in contact with the outer wall of the ø10 mm die. The sintering procedure was conducted at 1200 °C under 150–300 MPa of uniaxial pressure. The heating rate was 50 °C/min and the holding time at the highest temperature was 15 min. Grain growth heat treatments were performed in air for 8, 24 and 72 h at a temperature range of 1300–1450 °C. X-ray powder diffraction (XRD) was performed using a Rigaku RINT 2100 diffractometer with Cu Kα radiation (Tokyo, Japan). The operating parameters were 40 KV and 40 mA with a 2θ step of 0.02°. Cell parameters were calculated from the diffractions obtained using the MDI Jade 2010 software package (version 2.8.1, 2014, Materials Data, Livermore, CA, USA).

Microstructure was studied using high-resolution scanning electron microscopy (HRSEM, JEOL-7400F, Tokyo, Japan) and by transmission electron microscopy (TEM) using a JEOL 2100 (Tokyo, Japan) microscope equipped with a high-angle annular dark-field (HAADF) GATAN detector. Samples for scanning electron microscope (SEM) characterization were prepared using a standard metallographic procedure, finalized by polishing with a 1 μm diamond paste. Polished specimens were thermally etched at the same heat treatment temperature for 6 min.

TEM and STEM (scanning transmission electron microscope) samples were prepared from a copper-matrix composite with the spinel samples being embedded in the soft copper matrix, as described in detail by Halabi *et al.* [26] this technique was used in order to overcome charge-related issues encountered during the TEM work. The spinel samples were ground and mixed with pure copper powder (~10 μm). Disks 3 mm in diameter and 70 μm thick were pressed and sintered at 700 °C in an N_2 atmosphere. The perforation stage was carried out using a Gatan Dimpler and Precision Ion Polishing System. Grain size was estimated using Thixomet software [27] for image analysis. The density of the specimens was measured by the Archimedes method (ASTM Standard B-311 [28]), while Vickers hardness was measured using a Buehler–Micromet 2100 hardness tester (2 kg load, ASTM Standard C-1327 [29]). The samples were polished to an optical level for transmission measurements at 500 and 1000 nm wavelengths (Spectrophotometer V-1100D, MRC, Holon, Israel).

3. Results and Discussion

3.1. Phase Composition

Figure 1 shows XRD patterns for Li-doped and undoped nano-crystalline $MgO \cdot 1.06Al_2O_3$ and $MgO \cdot 1.21Al_2O_3$ samples synthesized by the combustion synthesis technique. The patterns indicate the presence of a spinel phase with relatively broad reflection peaks, suggesting small crystallite sizes calculated to range between 9.2 ± 0.2 and 32.5 ± 0.6 nm in the pure and doped samples (Table 1). Detailed characterization of the nano-powders prepared by this method can be found elsewhere [15].

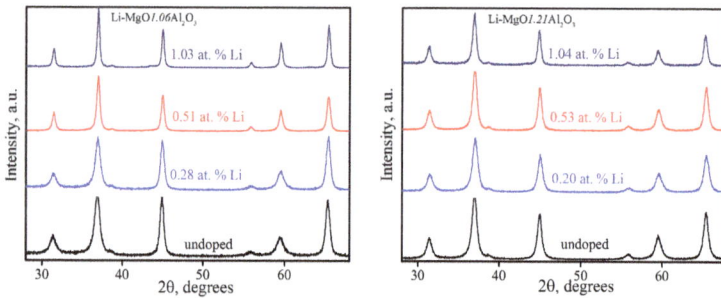

Figure 1. XRD patterns of as-synthesized powder samples.

Typical SPS-processed specimens from as-synthesized $MgO \cdot 1.06Al_2O_3$ powders containing different amounts of lithium are shown in Figure 2. The effect of lithium on the translucency of the $MgO \cdot 1.06Al_2O_3$ specimens is very apparent. In the present study, no attempts to determine optimal sintering conditions were made, with all of the compositions being sintered under the same conditions. The density, transmittance and hardness values (Table 1) of the samples prepared from near-stoichiometric powders ($n = 1.06$) all show maxima in the 0.51 at. % $Li-MgO \cdot 1.06Al_2O_3$ composition. Alumina-rich powders ($n = 1.21$) containing up to 0.53 at. % Li only reached about 95% of the theoretical density under these sintering conditions. Moreover, the samples showed no change in density, transmittance or hardness up to 0.53 at. % Li. At a higher lithium content (*i.e.*, 1.04 at. %), enhanced sinterability was observed.

0.00 at. % 0. 28at. % 0.51 at. % 1.03 at. %

Figure 2. Photograph of $Li-MgO \cdot 1.06Al_2O_3$ SPS-processed samples. The polished specimens are 10 mm in diameter and ~1.5 mm thick. The effect of lithium on transparency is visible.

Table 1. Cell parameter grain size, density, transmittence, hardness and MgO s.s. amount for SPS-processed Li-doped MgO· nAl$_2$O$_3$ (n = 1.06 and 1.21) samples.

Li (at. %)	A (Å)	D (nm)	ρ (g/cm³)	Trans. (500 nm) (%)	Trans. (1000 nm) (%)	Hardness (GPa)	Mg$_x$(Al,Li)$_{1-x}$O		Composition *	
							Wt. %	a (Å)	Mg	(Al, Li)
Li-MgO · 1.06Al$_2$O$_3$										
-	8.0810 (1)	102 ± 3	3.49 ± 0.01	-	-	14.3 ± 0.2	-	-	-	-
0.28 ± 0.02	8.0815 (1)	160 ± 5	3.54 ± 0.01	3.5 ± 0.1	14.4 ± 0.1	14.7 ± 0.3	0.9 ± 0.1	4.180 (9)	0.86	0.14
0.51 ± 0.05	8.0784 (1)	171 ± 3	3.56 ± 0.01	25.0 ± 0.1	45.3 ± 0.1	15.3 ± 0.4	1.8 ± 0.1	4.141 (9)	0.68	0.32
1.03 ± 0.10	8.0773 (1)	150 ± 8	3.54 ± 0.01	7.4 ± 0.1	22.9 ± 0.1	14.6 ± 0.5	2.9 ± 0.2	4.127 (9)	0.63	0.37
Li-MgO · 1.21Al$_2$O$_3$										
-	8.0647 (1)	81 ± 1	3.48 ± 0.01	-	-	14.4 ± 0.5	-	-	-	-
0.20 ± 0.02	8.0654 (1)	86 ± 2	3.49 ± 0.01	-	-	14.2 ± 0.2	-	-	-	-
0.53 ± 0.06	8.0656 (4)	94 ± 2	3.48 ± 0.01	-	-	14.1 ± 0.4	-	-	-	-
1.04 ± 0.10	8.0779 (2)	138 ± 4	3.61 ± 0.01	2.0 ± 0.1	10 ± 0.1	13.9 ± 0.3	2.3 ± 0.3	4.117 (9)	0.6	0.4

* Calculated using the Vegard rule and data from Doman's work [30].

The microstructures of the different SPS-processed specimens are presented in Figure 3. While the microstructure of the undoped MgO· 1.06Al$_2$O$_3$ sample displayed a homogeneous nano-structure with equiaxed grains (Figure 3), the Li-doped samples consisted of two grain size populations. The doped and undoped MgO· 1.21Al$_2$O$_3$ samples with lithium doping lower than 1.04 at. % seemed to be unaffected by the lithium addition and displayed similar equiaxed microstructures (Figure 3). The 1.04 at. % Li-MgO· 1.21Al$_2$O$_3$ sample, however, showed a similar microstructure to the 1.03 at. % Li-MgO· 1.06Al$_2$O$_3$ sample. The corresponding grain size distribution (an example is shown in Figure 4) exhibited a log-normal characteristic for all samples, with the calculated values summarized in Table 1. The grain size of near-stoichiometric specimens (n = 1.06) increased monotonically with the addition of lithium. However, this value appeared constant in alumina-rich powders (n = 1.21) containing up to 0.53 at. % Li. At higher lithium content (1.04 at. %), this value increased.

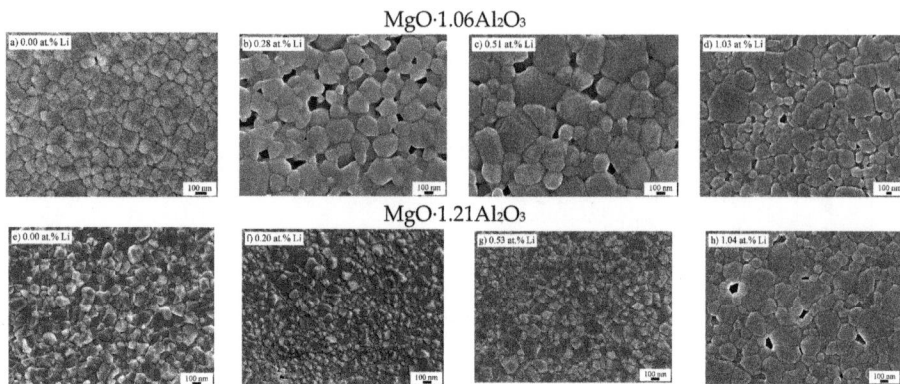

Figure 3. SEM micrographs of MgO· 1.06Al$_2$O$_3$ and MgO· 1.21Al$_2$O$_3$ SPS-processed samples. (**a**)–(**h**).

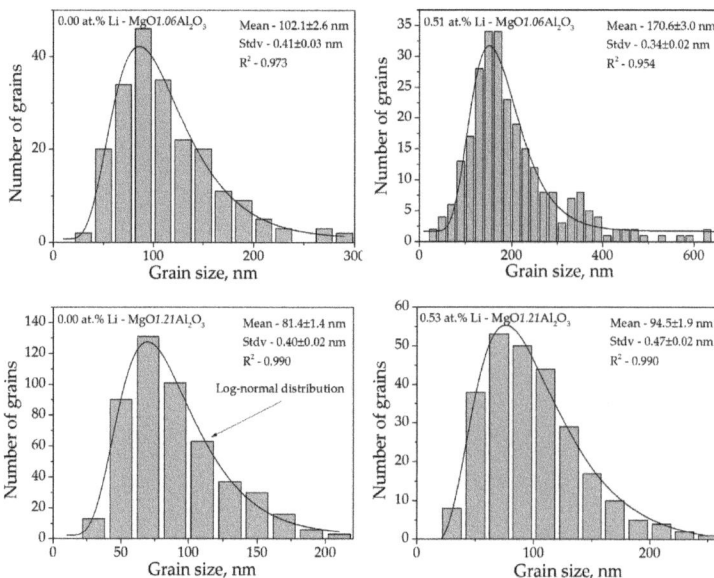

Figure 4. Typical grain size distribution and log-normal fitting of MgO· 1.06Al$_2$O$_3$ and MgO· 1.21Al$_2$O$_3$ with and without lithium addition processed by SPS.

The XRD patterns of the SPS-processed specimens are shown in Figure 5. The SPS-processed MgO· 1.06Al$_2$O$_3$ and 0.00–0.51 at. % Li-MgO· 1.21Al$_2$O$_3$ samples remained as a solid solution, while in the case of the 1.04 at. % Li-MgO· 1.21Al$_2$O$_3$ and 0.28 through 1.03 at. % Li-MgO· 1.06Al$_2$O$_3$ samples, Mg(Al,Li)O solid solution (MgO s.s.) and γ-LiAlO$_2$ [30,31] precipitated. The amounts of MgO s.s. precipitation were calculated using the Vegard rule and data from Reference [30] and are listed in Table 1. It should be noted that the γ-LiAlO$_2$ reflections were barely within the detection limit level of the XRD and were estimated to account for less than 1 wt. %. Similar behavior was found for the same powders after annealing at 1350 °C for 8 min in air [15].

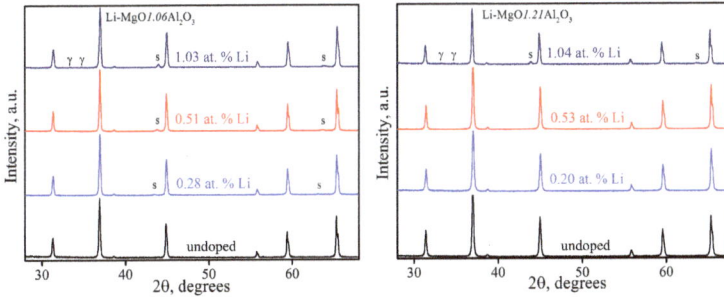

Figure 5. XRD patterns of Li-doped MgO· nAl$_2$O$_3$ (n = 1.06 and 1.21) SPS-processed samples. The precipitated MgO s.s. and γ-LiAlO$_2$ phases are marked by s and γ, respectively.

In general, for samples containing up to 53.0 at. % Li, the cell parameters were 8.0810 ± 0.0005 Å and 8.0652 ± 0.0005 Å for n = 1.06 and 1.21, respectively. At higher Li content, both 1.03 at. % Li-MgO· 1.06Al$_2$O$_3$ and 1.04 at. % Li-MgO· 1.21Al$_2$O$_3$ samples displayed the same cell parameter (8.0776 ± 0.0004 Å).

3.2. Grain Growth

The undoped, 0.28 and 0.53 at. % Li-doped MgO· 1.21Al$_2$O$_3$ SPS-processed samples remained as a solid solution, all the while exhibiting homogeneous microstructures with equiaxed polyhedral-shaped grains. To reveal the effect of lithium on grain growth mechanisms, the grain sizes resulting from a set of heat treatments at various temperatures and times were measured (Table 2, Figure 6).

Table 2. Grain sizes of heat-treated, undoped and 0.28 and 0.53 at. % Li-doped MgO·1.21Al$_2$O$_3$ samples.

Temperature (°C)/Time (h)	Grain Size (nm)		
	8	24	72
MgO· 1.21Al$_2$O$_3$			
1300	105 ± 8	125 ± 10	229 ± 6
1375	131 ± 10	200 ± 19	272 ± 26
1450	181 ± 12	292 ± 26	513 ± 22
0.28-MgO· 1.21Al$_2$O$_3$			
1300	111 ± 5	147 ± 10	157 ± 7
1375	154 ± 5	242 ± 18	249 ± 18
1450	306 ± 11	438 ± 9	991 ± 115
0.53-MgO· 1.21Al$_2$O$_3$			
1300	113 ± 8	116 ± 5	201 ± 7
1375	149 ± 8	237 ± 53	280 ± 8
1450	197 ± 18	300 ± 6	948 ± 47

1375 °C

| 0.00 at. %Li | 0.20 at. %Li | 0.53 at. %Li |

1450 °C

| 0.00 at. %Li | 0.20 at. %Li | 0.53 at. %Li |

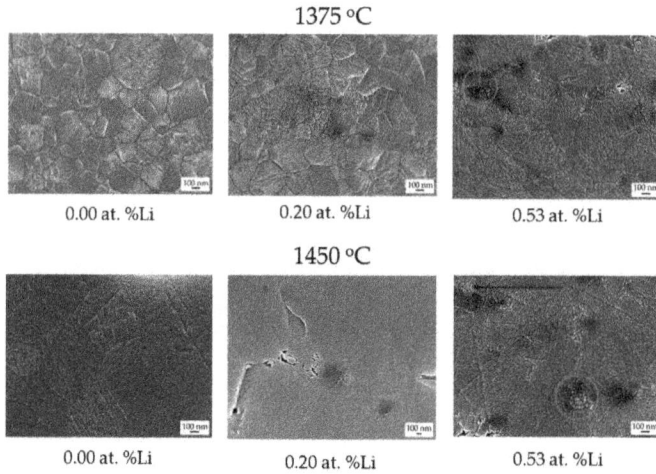

Figure 6. Micrographs of the undoped, 0.2 and 0.53 at. % Li-doped MgO· 1.21Al$_2$O$_3$ samples after heat-treatment at 1375 and 1450 °C for 24 h. The presence of the fine grain clusters is marked by red circles.

The undoped MgO· 1.21Al$_2$O$_3$ sample showed monotonic grain growth with temperature and time. The lithium-doped samples, however, presented a more complex behavior. At low temperatures and short holding times, the lithium-doped samples showed a monotonic-like behavior similar to the undoped samples. At higher temperatures (*i.e.*, 1450 °C, 8 h) or longer dwelling periods (*i.e.*, 1300 °C, 24 h), the 0.53 at. % Li-MgO· 1.21Al$_2$O$_3$ sample displayed lesser growth than the 0.20 at. % Li-MgO· 1.21Al$_2$O$_3$ sample (Figure 6). After a longer thermal exposure, namely 72 h at 1450 °C (Figure 6), the doped samples showed enhanced grain growth, reaching a size double that of the undoped sample.

Closer examination of the SEM images of the samples after heat treatment for 24 h at 1375 and 1450 °C (Figure 6) revealed the presence of small clusters of fine grains between larger grains in the doped samples. This finding suggests that lithium-rich phases may have precipitated during the heat treatments, which could explain the growth behavior of the doped samples.

Unfortunately, XRD analysis of these samples indicated only the presence of a spinel phase (Figure 7). Although no second phase was found, it might still be present, but it would remain undetected by the XRD technique if the phase only had a minor vol % and nano-sized dimensions [32].

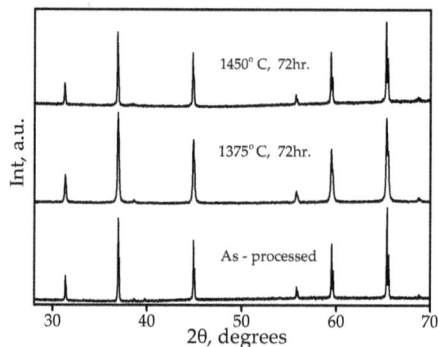

Figure 7. XRD spectra of as-processed and heat-treated to 1375 °C (72 h) and 1450 °C (72 h) samples.

To identify the nature of these fine grains, TEM analysis was performed on 0.20 at. % Li-MgO· 1.21Al$_2$O$_3$ before and after heat treatment at 1450 °C for 24 and 72 h (Figure 8). The TEM image of the SPS-processed 0.20 at. % Li-MgO· 1.21Al$_2$O$_3$ sample (Figure 8a) showed only spinel grains and confirmed the results of the XRD investigation regarding phase composition. After heat treatment at 1450 °C for 24 h, the presence of nano-particles of γ-lithium aluminate at the grain boundaries was detected (Figures 8b and 9).

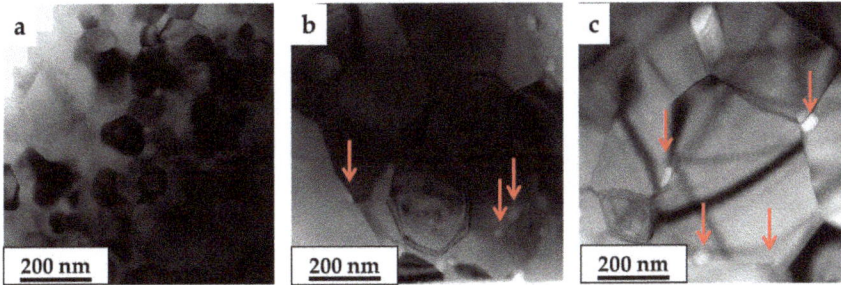

Figure 8. BF-TEM images of 0.20 at. % Li-MgO· 1.21Al$_2$O$_3$ spinel system as-processed (**a**); heat-treated at 1450 °C for 24 h (**b**); and heat-treated at 1450 °C for 72 h (**c**). Second phase in grain boundary, especially in triple points, is clearly visible in the heat-treated samples.

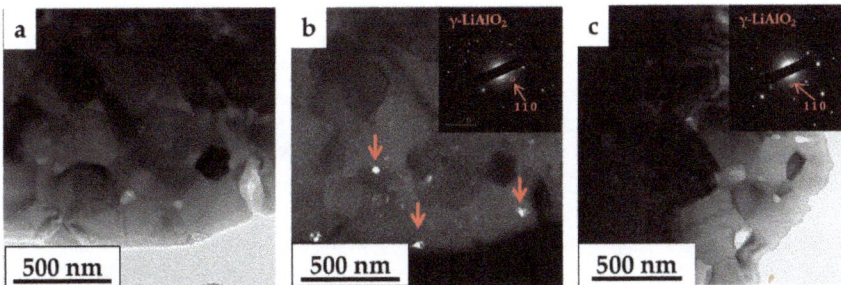

Figure 9. BF-TEM image of SPS spinel samples doped with 0.20 at. % Li after heat treatment at 1450 °C for 24 h (**a**); the white particles in the DF-TEM image are the γ-LiAlO$_2$ phase (**b**); BF-TEM image of spinel samples doped with 0.20 at. % Li after heat treatment at 1450 °C for 72 h (**c**); the selected area diffraction patterns indicate the reflection of (1 1 0) of the γ-LiAlO$_2$ phase.

In a previous study, we showed that the solubility limit of lithium in a spinel structure is controlled both by the Al-to-Mg ratio and by grain size [15]. Thus, even though no signs of second phase precipitation were present in the as-sintered 0.20 and 0.53 at. % Li-MgO· 1.21Al$_2$O$_3$ samples, additional grain growth would promote lithium segregation to the grain boundaries and precipitation of a second phase. The segregation of lithium to the grain boundary increases the grain growth rate by reducing the grain boundary energy [15]. On the other hand, second phase precipitation impedes grain growth via the Zener pinning mechanism [33–36]. Such behavior can be seen in Figure 10. The 0.2 at. % Li-MgO· 1.21Al$_2$O$_3$ spinel shows enhanced grain growth up to 24 h (<D> ~140 nm), after which time the growth is inhibited for a prolonged period of annealing due to second phase precipitation. In the more Li-rich samples (*i.e.*, 0.53 at. % Li), grain growth was inhibited at an early stage due to earlier second phase appearance. Further coarsening was related to precipitate coarsening followed by the grain coarsening [36].

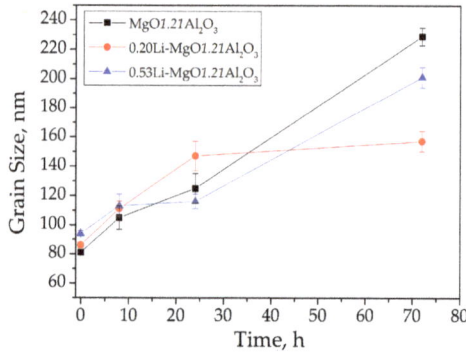

Figure 10. Grain sizes *vs.* annealing time of undoped, 0.2 and 0.53 at. % Li-doped $MgO \cdot 1.21Al_2O_3$ samples at 1300 °C.

Activation energy analysis of undoped, 0.2 and 0.53 at. % Li-doped $MgO \cdot 1.21Al_2O_3$ grain growth was performed using the phenomenological kinetic grain growth equation:

$$G_t^n - G_0^n = K_0 t exp \left(-\frac{Q}{RT} \right)$$

where G_t and G_0 are the grain sizes at times t and $t = 0$, respectively, n is the grain growth exponent, K_0 is the pre-exponential constant of the diffusion coefficient, Q is the activation energy for grain growth, T is the absolute temperature, and R is the gas constant.

The grain growth exponent or n value is readily determined as the inverse of the slope of a log G *vs.* log t plot. Using the original particle size as G_0, the grain size data can be fitted to linear lines with similar correlation factors ($R = 0.998$ and 0.937) for both the grain growth exponents of $n = 2$ (grain boundary–controlled diffusion) and $n = 3$ (lattice-controlled diffusion). This is in agreement with other works using either $n = 2$ or 3 [37,38]. Using $n = 2$, the activation energy and kinetic constant (K_0) for undoped $MgO \cdot 1.21Al_2O_3$ were found to be 288 ± 40 kJ·mol^{-1} and 2.09×10^6 μm^2/h. These values are in agreement with other data and are found between the values for $MgAl_2O_4$ and $MgO \cdot 1.56Al_2O_3$ (Table 3). The activation energies and K_0 for 0.2 and 0.53 at. % Li-$MgO \cdot 1.21Al_2O_3$ were found to be 670 ± 45, 543 ± 40 kJ·mol^{-1} and 3.41×10^{18}, 3.78×10^{14} μm^2/h, respectively; these values are significantly higher than those of the undoped sample. These findings are in line with the effect of the Zener pining mechanism, where grain growth is impeded at early stages by the secondary phase. Once the secondary phase has grown and the impediment is lifted, the spinel grains show enhanced growth (see data in Table 2) that can be attributed to the effect of lithium on the diffusion, by way of imposing oxygen vacancies [9–11,15].

Table 3. Grain growth parameters for 0–0.53 at. % Li-$MgO \cdot 1.21Al_2O_3$.

$MgO \cdot nAl_2O_3$	Activation Energy for Grain Growth (kJ/mol)	$ln(K_0)$
	Undoped	
1.56 (Chiang [39])	248 ± 29	16.35
1.21 (This study)	288 ± 40	14.55
1.013 (Chiang [39])	422 ± 10	28.23
~1.00 (Bratton [40])	462	30.54
at. % Li	Lithium doped $n = 1.21$ (This study)	
0.20	670 ± 45	42.67
0.53	543 ± 40	33.56

4. Summary

The effects of lithium doping on the sintering and grain growth kinetics of non-stoichiometric nano-MgO· nAl$_2$O$_3$ spinel with $n = 1.06$ and 1.21 were studied using a spark plasma sintering apparatus. The near-stoichiometry ($n = 1.06$) Li-doped samples showed higher sinterability in comparison with the aluminum-rich samples ($n = 1.21$) but also lower phase stability, with Mg(Li,Al)O and γ-LiAlO$_2$ phases precipitating during the course of the sintering process. Still, the aluminum-rich system ($n = 1.21$) showed greater phase stability up to 1 at. % of lithium for samples with grain sizes lower than 100 nm. The grain growth study indicated that in the Li-MgO· nAl$_2$O$_3$ system, grain growth was controlled by the Zener pining mechanism, where γ-LiAlO$_2$ precipitated at the grains boundaries. The activation energies of the undoped, 0.20 and 0.53 at. % Li-MgO· 1.21Al$_2$O$_3$ samples were 288 ± 40, 670 ± 45 and 543 ± 40 kJ· mol^{-1}, respectively.

Acknowledgments: This work was partially supported by the FP7-PEOPLE-2012-CIG (grant 321838-EEEF-GBE-CNS).

Author Contributions: Yuval Mordekovitz was responsible for the literature review, prepared samples for characterization, analyzed microstructural images and XRD patterns, and processed the experimental data. Lee Shelly performed the grain growth experiments, analyzed the grain growth data and interpreted the results. Mahdi Halabi prepared and executed the TEM experiments, analyzed the data and interpreted the results. Sergey Kalabukhov fabricated samples by SPS. Shmuel Hayun supervised the work and interpreted the results. All authors contributed to the writing of the paper.

Conflicts of Interest: The authors declare no conflict of interest.

Abbreviations

The following abbreviations are used in this manuscript:

SPS	spark plasma sintering
HRSEM	high resolution scanning electron microscope
HRTEM	high resolution transmission electron microscope
HAADF	high angle annular dark field
STEM	scanning transmission electron microscope
XRD	X-ray powder diffraction
MgO s.s.	Mg(Al,Li)O solid solution
BF-TEM	bright field transmission electron microscope
DF-TEM	dark field transmission electron microscope

References

1. Goldstein, A. Correlation between MgAl$_2$O$_4$-spinel structure, processing factors and functional properties of transparent parts (progress review). *J. Eur. Ceram. Soc.* **2012**, *32*, 2869–2886. [CrossRef]
2. Ganesh, I.; Teja, K.A.; Thiyagarajan, N.; Johnson, R.; Reddy, B.M. Formation and densification behavior of magnesium aluminate spinel: The influence of CaO and moisture in the precursors. *J. Am. Ceram. Soc.* **2005**, *88*, 2752–2761. [CrossRef]
3. Ganesh, I.; Bhattacharjee, S.; Saha, B.P.; Johnson, R.; Mahajan, Y.R. A new sintering aid for magnesium aluminate spinel. *Ceram. Int.* **2001**, *27*, 773–779. [CrossRef]
4. Huang, J.L.; Sun, S.Y.; Chen, C.Y. Investigation of high alumina-spinel: Effects of LiF and CaCO$_3$ addition (part 2). *Mater. Sci. Eng. A* **1999**, *259*, 1–7. [CrossRef]
5. Krell, A.; Hutzler, T.; Klimke, J.; Potthoff, A. Fine-grained transparent spinel windows by the processing of different nanopowders. *J. Am. Ceram. Soc.* **2010**, *93*, 2656–2666. [CrossRef]
6. Du Merac, M.R.; Reimanis, I.E.; Smith, C.; Kleebe, H.-J.; Müller, M.M. Effect of impurities and LiF additive in hot-pressed transparent magnesium aluminate spinel. *Int. J. Appl. Ceram. Technol.* **2013**, *10*, E33–E48. [CrossRef]
7. Chen, S.-K.; Cheng, M.-Y.; Lin, S.-J. Reducing the sintering temperature for MgO-Al$_2$O$_3$ mixtures by addition of cryolite (Na$_3$AlF$_6$). *J. Am. Ceram. Soc.* **2004**, *85*, 540–544. [CrossRef]

8. Bhattacharya, G.; Zhang, S.; Smith, M.E.; Jayaseelan, D.D.; Lee, W.E. Mineralizing magnesium aluminate spinel formation with B_2O_3. *J. Am. Ceram. Soc.* **2006**, *89*, 3034–3042. [CrossRef]
9. Rozenburg, K.; Reimanis, I.E.; Kleebe, H.-J.; Cook, R.L. Chemical interaction between LiF and $MgAl_2O_4$ spinel during sintering. *J. Am. Ceram. Soc.* **2007**, *90*, 2038–2042. [CrossRef]
10. Meir, S.; Kalabukhov, S.; Froumin, N.; Dariel, M.P.; Frage, N. Synthesis and densification of transparent magnesium aluminate spinel by SPS processing. *J. Am. Ceram. Soc.* **2009**, *92*, 358–364. [CrossRef]
11. Rozenburg, K.; Reimanis, I.E.; Kleebe, H.-J.; Cook, R.L. Sintering kinetics of a $MgAl_2O_4$ spinel doped with LiF. *J. Am. Ceram. Soc.* **2008**, *91*, 444–450. [CrossRef]
12. Du Merac, M.R.; Kleebe, H.-J.; Müller, M.M.; Reimanis, I.E. Fifty years of research and development coming to fruition; Unraveling the complex interactions during processing of transparent magnesium aluminate ($MgAl_2O_4$) spinel. *J. Am. Ceram. Soc.* **2013**, *96*, 3341–3365. [CrossRef]
13. Carnall, E. The densification of MgO in the presence of a liquid phase. *Mater. Res. Bull.* **1967**, *2*, 1075–1086. [CrossRef]
14. Reimanis, I.; Kleebe, H.-J. A review on the sintering and microstructure development of transparent spinel ($MgAl_2O_4$). *J. Am. Ceram. Soc.* **2009**, *92*, 1472–1480. [CrossRef]
15. Mordekovitz, Y.; Hayun, S. On the effect of lithium on the energetics and thermal stability of nano-sized nonstoichiometric magnesium aluminate spinel. *J. Am. Ceram. Soc.* **2016**. [CrossRef]
16. Morita, K.; Kim, B.-N.; Hiraga, K.; Yoshida, H. Fabrication of transparent $MgAl_2O_4$ spinel polycrystal by spark plasma sintering processing. *Scr. Mater.* **2008**, *58*, 1114–1117. [CrossRef]
17. Morita, K.; Kim, B.-N.; Yoshida, H.; Hiraga, K. Densification behavior of a fine-grained $MgAl_2O_4$ spinel during spark plasma sintering (SPS). *Scr. Mater.* **2010**, *63*, 565–568. [CrossRef]
18. Frage, N.; Cohen, S.; Meir, S.; Kalabukhov, S.; Dariel, M.P. Spark plasma sintering (SPS) of transparent magnesium-aluminate spinel. *J. Mater. Sci.* **2007**, *42*, 3273–3275. [CrossRef]
19. Morita, K.; Kim, B.-N.; Yoshida, H.; Hiraga, K. Spark-plasma-sintering condition optimization for producing transparent $MgAl_2O_4$ spinel polycrystal. *J. Am. Ceram. Soc.* **2009**, *92*, 1208–1216. [CrossRef]
20. Rothman, A.; Kalabukhov, S.; Sverdlov, N.; Dariel, M.P.; Frage, N. The effect of grain size on the mechanical and optical properties of spark plasma sintering-processed magnesium aluminate spinel $MgAl_2O_4$. *Int. J. Appl. Ceram. Technol.* **2014**, *11*, 146–153. [CrossRef]
21. Bonnefont, G.; Fantozzi, G.; Trombert, S.; Bonneau, L. Fine-grained transparent $MgAl_2O_4$ spinel obtained by spark plasma sintering of commercially available nanopowders. *Ceram. Int.* **2012**, *38*, 131–140. [CrossRef]
22. Fu, P.; Lu, W.; Lei, W.; Xu, Y.; Wang, X.; Wu, J. Transparent polycrystalline $MgAl_2O_4$ ceramic fabricated by spark plasma sintering: Microwave dielectric and optical properties. *Ceram. Int.* **2013**, *39*, 2481–2487. [CrossRef]
23. Morita, K.; Kim, B.-N.; Yoshida, H.; Zhang, H.; Hiraga, K.; Sakka, Y. Effect of loading schedule on densification of $MgAl_2O_4$ spinel during spark plasma sintering (SPS) processing. *J. Eur. Ceram. Soc.* **2012**, *32*, 2303–2309. [CrossRef]
24. Khasanov, O.; Dvilis, E.; Khasanov, A.; Polisadova, E.; Kachaev, A. Optical and mechanical properties of transparent polycrystalline $MgAl_2O_4$ spinel depending on SPS conditions. *Phys. Status Solidi* **2013**, *10*, 918–920. [CrossRef]
25. Ianoş, R.; Lazău, I.; Păcurariu, C.; Barvinschi, P. Solution combustion synthesis of $MgAl_2O_4$ using fuel mixtures. *Mater. Res. Bull.* **2008**, *43*, 3408–3415. [CrossRef]
26. Halabi, M.; (Ben-Gurion University of the Negev, Beer-Sheva, Israel). Personal communication, 2016.
27. Hayun, S.; Dilman, H.; Dariel, M.P.; Frage, N.; Dub, S. Effect of the carbon source on the microstructure and mechanical properties of reaction bonded boron carbide. In *Ceramic Transactions 209 (Advances in Sintering Science and Technology)*; American Ceramic Society: Westerville, OH, USA, 2010; pp. 29–41.
28. ASTM B-311. Standard test method for density of powder metallurgy (PM) materials containing less than two percent porosity 1. *ASTM Int.* **2008**. [CrossRef]
29. C-1327 A. Standard test method for vickers indentation hardness of advanced ceramics 1. *ASTM Int.* **2008**. [CrossRef]
30. Doman, R.C.; McNally, R.N. Solid solution studies in the MgO-LiAlO system. *J. Mater. Sci.* **1973**, *8*, 189–191. [CrossRef]
31. Izquierdo, G.; West, A.R. Compatibility relations in the system Li_2O-MgO-Al_2O_3. *J. Am. Ceram. Soc.* **1980**. [CrossRef]

32. Garitaonandia, J.S.; Gorria, P.; Barquín, L.F.; Barandiarán, J.M.; Barqu'in, L.F.; Barandiarán, J.M. Low-temperature magnetic properties of Fe nanograins in an amorphous Fe-Zr-B matrix. *Phys. Rev. B* **2000**, *61*, 6150–6155. [CrossRef]

33. Hassold, G.N.; Holm, E. Effects of particle size on inhibited grain growth. *Scr. Met. Mater.* **1990**, *24*, 11–13. [CrossRef]

34. Grewal, G.; Ankem, S. Modeling matrix grain growth in the presence of growing second phase particles in two phase alloys. *Acta Metall. Mater.* **1990**, *38*, 1607–1617. [CrossRef]

35. Fan, D.; Chen, L.-Q.; Chen, S.-P.P. Numerical simulation of zener pinning with growing second-phase particles. *J. Am. Ceram. Soc.* **1998**, *81*, 526–532. [CrossRef]

36. Rios, P.R. Overview no. 62. A theory for grain boundary pinning by particles. *Acta Metall.* **1987**, *35*, 2805–2814. [CrossRef]

37. Lappalainen, R.; Pannikkat, A.; Raj, R. Superplastic flow in a non-stoichiometric ceramic: Magnesium aluminate spinel. *Acta Metall. Mater.* **1993**, *41*, 1229–1235. [CrossRef]

38. Chaim, R. Activation energy and grain growth in nanocrystalline Y-TZP ceramics. *Mater. Sci. Eng. A* **2008**, *486*, 439–446. [CrossRef]

39. Chiang, Y. Grain boundary mobility and segregation in non-stoichiometric solid solutions of magnesium aluminate spinel. *J. Am. Ceram. Soc.* **1980**. [CrossRef]

40. Bratton, R.J. Sintering and grain-growth kinetics of something is missing here. *J. Eur. Ceram. Soc.* **1970**, *35*, 141–143.

materials

MDPI

Article

Creep of Polycrystalline Magnesium Aluminate Spinel Studied by an SPS Apparatus

Barak Ratzker, Maxim Sokol, Sergey Kalabukhov and Nachum Frage *

Department of Materials Engineering, Ben-Gurion University of the Negev, P.O. Box 653, Beer-Sheva 84105, Israel; ratzkerb@post.bgu.ac.il (B.R.); sokolmax@bgu.ac.il (M.S.); kalabukh@bgu.ac.il (S.K.)
* Correspondence: nfrage@bgu.ac.il; Tel.: +972-8-646-1468

Academic Editor: Eugene A. Olevsky
Received: 8 May 2016; Accepted: 16 June 2016; Published: 20 June 2016

Abstract: A spark plasma sintering (SPS) apparatus was used for the first time as an analytical testing tool for studying creep in ceramics at elevated temperatures. Compression creep experiments on a fine-grained (250 nm) polycrystalline magnesium aluminate spinel were successfully performed in the 1100–1200 °C temperature range, under an applied stress of 120–200 MPa. It was found that the stress exponent and activation energy depended on temperature and applied stress, respectively. The deformed samples were characterized by high resolution scanning electron microscope (HRSEM) and high resolution transmission electron microscope (HRTEM). The results indicate that the creep mechanism was related to grain boundary sliding, accommodated by dislocation slip and climb. The experimental results, extrapolated to higher temperatures and lower stresses, were in good agreement with data reported in the literature.

Keywords: creep; spinel; SPS

1. Introduction

The use of spark plasma sintering (SPS) has continuously expanded over the past 20 years thanks to its excellent sintering capabilities. A special configuration of SPS tooling, described in our previous work [1,2], makes it possible to apply uniaxial pressure up to 1 GP during the sintering process. Sintering under high pressure allows for significant reduction of processing temperatures and fabrication of nano-structured ceramics. It was reported [1] that nano-structured magnesium aluminate spinel specimens, possessing a unique combination of optical and mechanical properties, could be fabricated under a uniaxial pressure of 400 MPa at 1200 °C. One of the main densification mechanisms acting in the final stage of sintering under elevated pressure is high temperature deformation (creep) of ceramic particles or grains [3]. The creep behavior of polycrystalline magnesium aluminate spinel and its capability to undergo superplastic deformation at relatively high temperatures (1300–1800 °C) over a wide range of applied stresses (1–200 MPa) have been investigated [4–11]. However, to the best of our knowledge, there is no data on creep behavior under conditions close to those that are applied during the high pressure SPS process. The data output of the SPS system includes temperature, applied pressure, relative punch displacement (RPD) and electric pulse parameters (*i.e.*, voltage, mode of current, frequency, *etc.*). In principle, the SPS apparatus is a high temperature dilatometer and can be used for the investigation of mechanical properties of ceramics at high temperatures. The accuracy of RPD measurements (about 1 μm) is suitable for high temperature experiments, such as creep. In the present study, an SPS apparatus was used to investigate the creep behavior of magnesium aluminate spinel for the first time.

2. Materials and Experimental Procedures

Magnesium aluminate spinel samples intended for creep testing were fabricated by SPS (HP-D10, FCT Systems, Rauensein, Germany) from a commercial MgO· Al_2O_3 powder (S30CR Baikowski Chimie, La Blame de Silingy, France) with a specific surface area of 30 m^2/g, impurities levels of 10, 10, 20 and 5 ppm for Fe, Na, Si and Ca, respectively. SPS was performed inside a graphite die with 12 mm height and 20 mm diameter. The sintering parameters were a sintering temperature of 1300 °C, a dwell time of 20 min, a heating rate of 10 °C/min, an applied pressure of 60 MPa and a cooling rate of 50 °C/min. The sintered samples were fully dense, with a grain size of about 250 nm. The spinel samples were precisely machined into a cylindrical geometry 12 mm in height and 6 mm in diameter. The creep experiments were conducted in the SPS apparatus with a DC pulse-mode current pattern (pulse 5 ms and pause 2 ms). A non-constrained sample was placed inside the high pressure SPS tooling, which consisted of an outer graphite die (outer diameter, 50 mm) and an inner die made of silicon carbide. The silicon carbide die had 10 mm inner and 20 mm outer diameters. A K-type thermocouple was inserted through the outer graphite die and placed in contact with the inner SiC die. Samples were heated to the initial temperature of 1100 °C at a heating rate of 200 °C/min. Each compression creep experiment was conducted under constant pressures of 120, 150 and 200 MPa at various temperatures of 1100, 1150 and 1200 °C, with a dwell time of about 2 h. The microstructure of the polished and thermally etched (1400 °C for 10 min under ambient atmosphere) samples was examined using a high resolution scanning electron microscope (HRSEM; JSM-7400, JEOL, Tokyo, Japan) The grain size was estimated by Thixomet software (Thixomet, St.-Petersburg, Russia) for image analysis [12,13]. Samples for high resolution transmission electron microscope (HRTEM) analysis were prepared by a focused ion beam (FIB; Helios NanoLab 600, FEI, Hillsboro, OR, USA) and examined using a high resolution transmission electron microscope (HRTEM; JEM-2010F, JEOL, Tokyo, Japan).

3. Results and Discussion

3.1. Strain Rate

Creep curves for spinel were obtained from the experimental data recorded by the SPS system. RPD was converted to strain according to the initial specimen height (Figure 1).

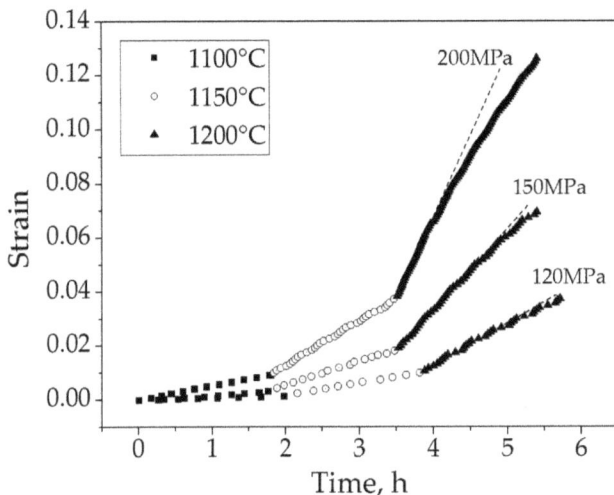

Figure 1. Creep curves for spinel under pressures of 120–200 MPa in the 1100–1200 °C temperature range. The dashed lines indicate change of the slope.

Each creep curve consists of three domains, each related to the creep at various temperatures, and which changed during the course of the experiment. At low temperatures (reflected as the first domain of the creep curves), the initial deformation was extremely low and almost undetectable. For the other curve domains, the material was already in the steady stage of creep. The softening of the material was clearly observed as an increased slope of the curves with temperature. Finally, the hardening effect can be observed at high temperature and upon high strain (reflected in the continuous change/decrease of the slopes and indicated by the dashed line in Figure 1). Strain rates were determined from the slopes of the quasi-linear steady-state portions [4,8,11] of each domain and are presented as a function of temperature in Figure 2.

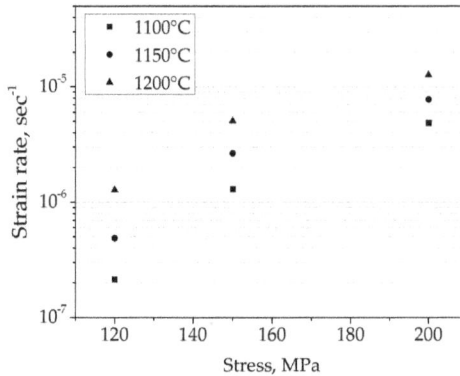

Figure 2. Creep rates of spinel as a function of pressure, tested at various temperatures.

The creep rates obtained at the tested temperatures and stress range fit the deformation-mechanism map for magnesium aluminate spinel [14] and correspond to the region of power law creep [15]. Therefore, the data were analyzed according to the general creep equation:

$$\dot{\varepsilon} = A\sigma^n \exp\left(\frac{-Q}{RT}\right) \tag{1}$$

where $\dot{\varepsilon}$ is the creep rate, A is the creep constant, σ is the applied load, n is the stress exponent, Q is the activation energy, R is the gas constant and T is the temperature. The creep parameters (*i.e.*, Q and n) can vary with experimental conditions (*i.e.*, applied stress and temperature), along with other factors, such as composition and microstructure of the tested materials [16]. The experimental data for different temperatures and stresses allow for estimations of stress exponent and apparent activation energy values according to:

$$\ln(\dot{\varepsilon}) = \ln A + n\ln(\sigma) - \frac{Q}{RT} \tag{2}$$

3.2. Stress Exponent

The values of the stress exponent (Table 1) were determined from the slopes of the curves presented in Figure 3.

Table 1. Values of the stress exponent at various temperatures.

Temperature (°C)	Stress Exponent (*n*)
1100	3.48 ± 0.1
1150	2.64 ± 0.26
1200	1.87 ± 0.15

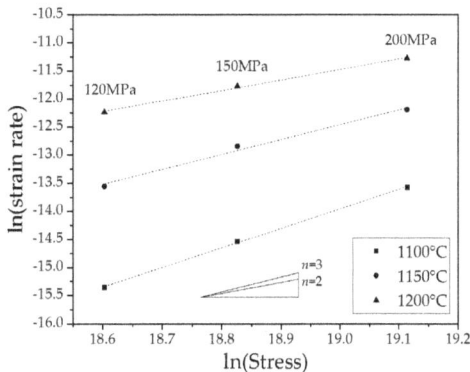

Figure 3. ln(strain rate) *vs.* ln(stress) for spinel tested under 120, 150 and 200 MPa.

The stress exponent values obtained are in good agreement with those previously reported [4–11]. Temperature-dependence of the stress exponent was also observed [9] and was attributed to changes in the creep mechanism. For low stress exponent values ($n \approx 1$), creep is governed by the diffusional flow of ions [17]. For higher stress exponent values ($n \approx 2$), creep of the fine polycrystalline ceramics is mostly controlled by grain boundary sliding (GBS) [18], as was clearly demonstrated for polycrystalline alumina [19,20]. GBS, however, has to be accommodated by an additional process [21,22], such as intergranular slip and the subsequent climb of dislocations [23]. According to the apparent values of the stress exponent, GBS was the dominant mechanism in the 1150–1200 °C temperature range. The higher stress exponent ($n \approx 3.5$) obtained at the lowest temperature considered (1100 °C), could be attributed to formation of triple-point folds [24,25]. Another possibility is the reduced contribution of GBS [26], with a larger percentage of deformation being carried out by dislocation slip and climb ($4 > n > 3$) [15].

3.3. Activation Energy

The temperature-dependence of the creep rate under various stresses allowed for estimation of the apparent activation energy of the process (Figure 4).

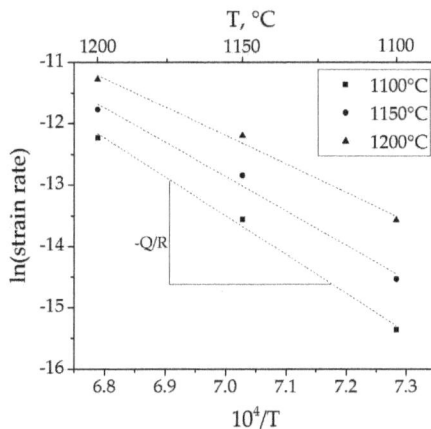

Figure 4. Strain rate *vs.* the reciprocal of temperature.

Calculated apparent activation energy values are presented in Table 2.

Table 2. Apparent activation energies for polycrystalline magnesium aluminate spinel.

Applied Stress (Mpa)	Activation Energy (Q) (kJ/mol)
120	526 ± 35
150	465 ± 50
200	387 ± 36

The apparent activation energy values for creep in spinel under our experimental conditions, as well as their dependence on the stress applied (Figure 5), are in good agreement with previously reported data [4,5,7]. The decrease in activation energy with applied stress may be attributed to competition between the GBS and dislocation creep mechanisms of deformation. Under higher applied stress, more dislocations are generated within the material, a process which requires relatively lower thermal activation, making the effect of temperature less pronounced. At lower applied stress, the amount of dislocations decreases, while the role of ion diffusion increases. As such, the effect of temperature becomes more significant.

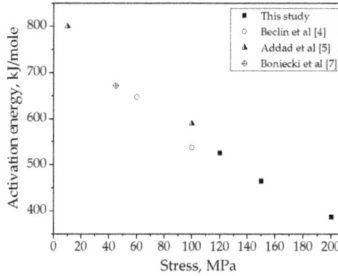

Figure 5. Apparent activation energy as a function of stress.

The average creep constant A (=8.43 × 10^{-11} ± 1.38 × 10^{-11} sec^{-1}· MPa^{-n}), which depends on the material and only slightly on pressure and temperature was estimated based on the stress exponent and activation energy values obtained. The experimental results were extrapolated for a wider range of strain rates (*i.e.*, for lower stresses and higher temperatures), to fit with experimental data reported in the literature [4–6]. The extrapolated data, along with our experimental results and the data reported in literature, are presented in Figure 6.

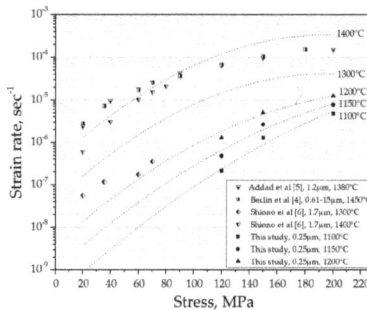

Figure 6. Strain rate *vs.* applied stress. Extrapolation of our experimental data (dashed lines) and values reported in the literature for various creep conditions and grain sizes (indicated in the legend) are shown.

The extrapolated values of the strain rate are in a good agreement with previously reported data. Some scattering of the experimental results may be attributed to differences in creep conditions, particularly the grain sizes of the tested samples.

It should be noted that creep was tested in the SPS apparatus under a small electric field ($\sim$70 V$\cdot$cm^{-1}). The good agreement noted between our results and those reported obtained using a conventional testing procedure indicates that the electric field contributes only a negligible or no effect on the creep of the fine grain polycrystalline spinel.

HRSEM images of polished and thermally etched samples before and after creep deformation are presented in Figure 7.

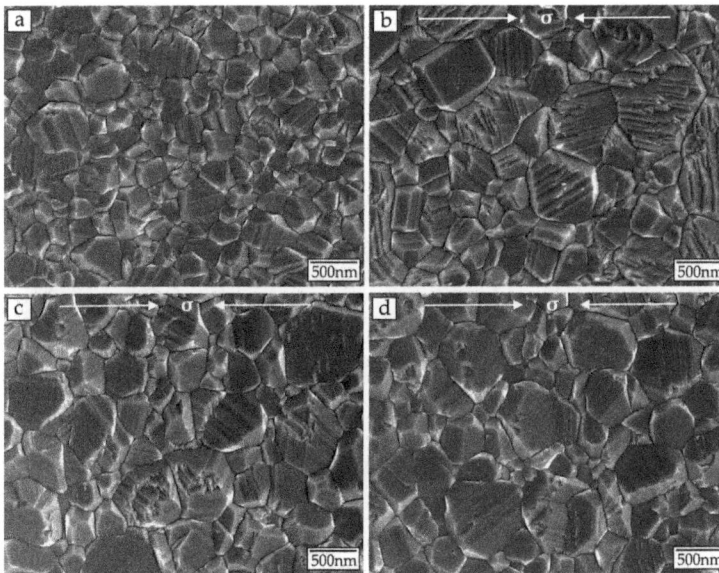

Figure 7. High resolution scanning electron microscope (HRSEM) images of the spinel samples: before creep (**a**); after creep at 1100–1200 °C (4% strain) under 120 (**b**); (7% strain) 150 (**c**) and (13% strain) 200 MPa (**d**). Compression direction is marked.

The initial microstructure (Figure 7a) consisted of fine equiaxed grains with an average size of about 250 nm. After the creep test, the average grain size was larger (400 nm) and grain shape remained equiaxed with no directional growth (Figure 7b–d). According to [27], this implies that GBS was the main mechanism by which the samples were deformed during creep under our testing conditions. We assume that in this range of grain size its effect on creep rate was not significant.

The appearance of faceting morphology of spinel grains is attributed to the preferential thermal etching of {111} planes, which have higher surface energy than either the {100} or {110} planes. The faceting morphology was detected in both before and after creep samples and does not depend on the grain deformation [28].

To clarify the creep mechanism, HRTEM analysis of deformed samples after creep under pressures of 120 and 200 MPa, was performed (Figure 8).

Several types of distinctive features of creep were found in all examined samples. The formation of triple point voids and displaced triple points (Figure 8a,b) provided strong evidence for GBS [29]. Moreover, grain separation, appeared as grain boundary cavities, was also observed, which is typical for creep with no glassy phase at the grain boundaries [25]. These observations confirmed that substantial GBS had taken place during creep under our testing conditions. Dislocations and dislocation pile-ups were also observed, especially in that sample subjected to 200 MPa pressure (Figure 8c). By using the weak-beam dark field (WBDF) method that allows pronounced contrast for dislocations [30], high dislocation densities were found within the grains at $g = 440$ (Figure 8d), which is known to be the preferred dislocation slip plane in spinel [11,30,31]. This may indicate that a dislocation mechanism was involved at the relatively higher stress level.

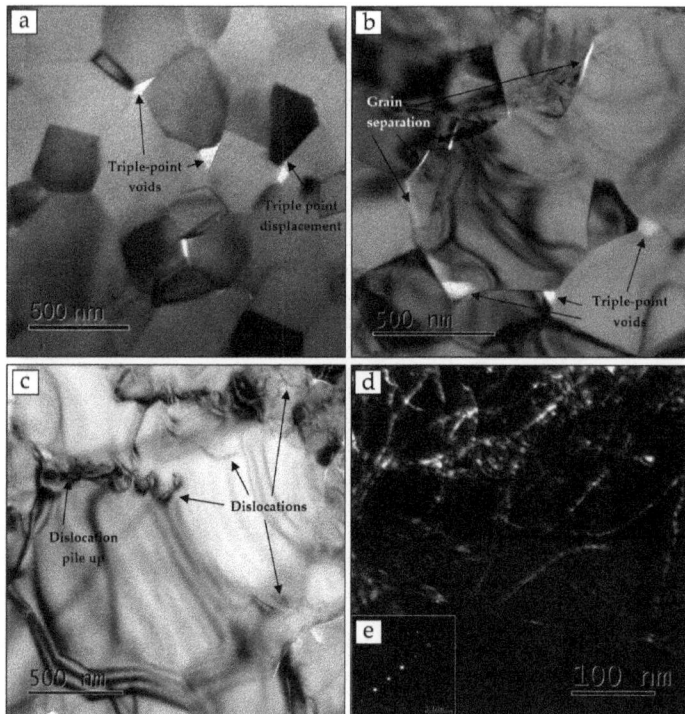

Figure 8. High resolution transmission electron microscope (HRTEM) images of spinel samples after deformation at 1100–1200 °C. Under 120 MPa (4% strain) triple-point voids and displaced triple points are shown (**a**); under 200 MPa (13% strain) grain separation and sliding along the grain boundaries (**b**) and dislocations (**c**) are shown. A weak-beam dark field (WBDF) image for $g = 440$ shows the high dislocation density within the grain after creep in response to 200 MPa pressure (**d**); The selected area diffraction pattern is presented in the insert (**e**). The examined cross-sections were perpendicular to the compression axis.

4. Conclusions

Compression creep testing of polycrystalline magnesium aluminate spinel was successfully performed for the first time using an SPS apparatus. The stress exponent value decreased from 3.5 to 2 with temperature in the 1100–1200 °C range. Likewise, the apparent activation energy value decreased from 526 to 387 kJ/mol with applied stress in the 120–200 MPa range. Our results and the data predicted by extrapolation are in a good agreement with values reported in the literature. The microstructure of the deformed samples, consisting of an equiaxed grain structure, indicated that GBS operated during creep. HRTEM examination confirmed that the deformation occurred by GBS accommodated by dislocation movement.

Materials **2016**, *9*, 493

Author Contributions: Barak Ratzker was responsible for literature review, prepared samples for characterization, analyzed microstructural images, and processed the experimental data. Maxim Sokol designed the experiments, prepared samples for creep testing, analyzed the creep data and interpreted the results. Sergey Kalabukhov fabricated samples by SPS and conducted creep experiments. Nachum Frage supervised the work and interpreted the results. All authors contributed to the writing of the paper.

Conflicts of Interest: The authors declare no conflict of interest.

References

1. Sokol, M.; Kalabukhov, S.; Dariel, M.P.; Frage, N. High-pressure spark plasma sintering (SPS) of transparent polycrystalline magnesium aluminate spinel (PMAS). *J. Eur. Ceram. Soc.* **2014**, *34*, 4305–4310. [CrossRef]
2. Sokol, M.; Kalabukhov, S.; Kasiyan, V.; Dariel, M.P.; Frage, N. Functional properties of Nd:YAG polycrystalline ceramics processed by High-Pressure Spark Plasma Sintering (HPSPS). *J. Am. Ceram. Soc.* **2016**, *99*, 802–807. [CrossRef]
3. Wilkinson, D.S.; Ashby, M.F. Pressure sintering by power law creep. *Acta Metall.* **1975**, *23*, 1277–1285. [CrossRef]
4. Beclin, F.; Duclos, R.; Crampon, J.; Valin, F. Microstructural superplastic deformation in MgO· Al$_2$O$_3$ spinel. *Acta Metall. Mater.* **1995**, *43*, 2753–2760. [CrossRef]
5. Addad, A.; Beclin, F.; Crampon, J.; Duclos, R. High temperature deformation of spinel-zirconia composites effect of zirconia content. *Ceram. Int.* **2007**, *33*, 1057–1063. [CrossRef]
6. Shiono, T.; Ishitomi, H.; Okamoto, Y.; Nishida, T. Deformation mechanism of fine-grained magnesium aluminate spinel prepared using and alkoxide precursor. *J. Am. Ceram. Soc.* **2000**, *83*, 645–647. [CrossRef]
7. Boniecki, M.; Librant, Z.; Wajler, A.; Wesolowski, W.; Weglarz, H. Fracture toughness, strength and creep of transparent ceramics at high temperature. *Ceram. Int.* **2012**, *38*, 4517–4524. [CrossRef]
8. Beclin, F.; Duclos, R.; Crampon, J.; Valin, F. Superplasticity of HIP MgO· Al$_2$O$_3$ spinel: Prospects for superplastic forming. *J. Eur. Ceram. Soc.* **1997**, *17*, 439–445. [CrossRef]
9. Choi, M.D.; Palmou, H., III. *Flow and Fracture of Hot-Pressed Polycrystalline Spinel at Elevated Temperatures*; No. TR-2; Dept. of Engineering Research, North Carolina State University: Raleigh, NC, USA, 1965.
10. Panda, P.C.; Raj, R.; Morgan, P.E.D. Superplastic deformation in fine-grained MgO· Al$_2$O$_3$ spinel. *J. Am. Ceram. Soc.* **1985**, *68*, 522–529. [CrossRef]
11. Morita, K.; Hiraga, K.; Kim, B.; Suzuki, T.S.; Sakka, Y. Strain softening and hardening during superplastic-like flow in a fine-grained MgAl$_2$O$_4$ spinel polycrystal. *J. Am. Ceram. Soc.* **2004**, *87*, 1102–1109. [CrossRef]
12. Kazakov, A.; Luong, N.; Kazakova, E.; Zorina, E. Thixomet image analyzer for characterization of 2D and 3D materials structure. *Microstruct. Sci.* **2000**, *27*, 133–142.
13. Kazakov, A.; Luong, N. Characterization of semisolid materials structure. *Mater. Charact.* **2001**, *46*, 155–161. [CrossRef]
14. Frost, H.J.; Ashby, M.F. *Deformation-Mechanism Maps: The Plasticity and Creep of Metals and Ceramics*; Pergamon Press: Oxford, UK, 1982; pp. 105–110.
15. Weertman, J. Theory of steady-state creep based on dislocation climb. *J. Appl. Phys.* **1955**, *26*, 1213–1217. [CrossRef]
16. Barsoum, M.W. *Fundamentals of Ceramics*; IOP Publishing Ltd.: Bristol, UK, 2003; pp. 400–432.
17. Chokshi, A.H. Diffusion creep in oxide ceramics. *J. Eur. Ceram. Soc.* **2002**, *22*, 2469–2478. [CrossRef]
18. Langdon, T.G.R. The role of grain boundaries in high temperature deformation. *J. Mater. Sci. Eng.* **1993**, *166*, 67–79. [CrossRef]
19. Cannon, W.R.; Sherby, O.D. Creep behavior and grain boundary sliding in polycrystalline Al$_2$O$_3$. *J. Am. Ceram. Soc.* **1977**, *60*, 44–47. [CrossRef]
20. Ruano, O.A.; Wadsworth, J.; Sherby, O.D. Deformation of fine-grained alumina by grain boundary sliding accommodated by slip. *Acta Mater.* **2003**, *51*, 3617–3634. [CrossRef]
21. Langdon, T.G. Grain boundary sliding as a deformation mechanism during creep. *Philos. Mag.* **1970**, *22*, 689–700. [CrossRef]
22. Raj, R.; Ashby, M.F. On grain boundary sliding and diffusional creep. *Metall. Trans.* **1971**, *2*, 1113–1127. [CrossRef]
23. Ball, A.; Hutchison, M.M. Superplasticity in the aluminium-zinc eutectoid. *Met. Sci.* **1969**, *3*, 1–7. [CrossRef]

24. Gifkins, R.C. Grain boundary sliding and its accommodation during creep and superplasticity. *Metall. Trans.* **1976**, *7*, 1225–1232. [CrossRef]

25. Cannon, W.R.; Langdon, T.G. Review creep of ceramics part 1 mechanical characteristics. *J. Mater. Sci.* **1983**, *18*, 1–50.

26. Cannon, W.R.; Langdon, T.G. Review creep of ceramics part 2 an examination of flow mechanisms. *J. Mater. Sci.* **1988**, *23*, 1–20. [CrossRef]

27. Langdon, T.G. Grain-boundary sliding in ceramics. *J. Am. Ceram. Soc.* **1972**, *55*, 430–431. [CrossRef]

28. Browne, D.; Li, H.; Giorgi, E.; Dutta, S.; Biser, J.; Vinci, R.P.; Chan, H.M. Templated epitaxial coatings on magnesium aluminate spinel using sol gel method. *J. Mater. Sci.* **2009**, *44*, 1180–1186. [CrossRef]

29. Heuer, A.H.; Cannon, R.M.; Tighe, N.J. Plastic deformation in fine-grain ceramics. In Proceedings of the 15th Sagamore Army Materials Research Conference, Ultrafine-Grain Ceramics, Raquette Lake, NY, USA, 20–23 August 1968; Burke, J.J., Reed, N.L., Weiss, V., Eds.; Syracuse University Press: New York, NY, USA, 1970; pp. 339–365.

30. Ting, C.J.; Lu, H.Y. Hot-pressing of magnesium aluminate spinel—II. Microstructure development. *Acta Mater.* **1999**, *47*, 831–840. [CrossRef]

31. Mitchell, T.E. Dislocations and mechanical properties of $MgO\text{-}Al_2O_3$ spinel single crystals. *J. Am. Ceram. Soc.* **1999**, *82*, 3305–3316. [CrossRef]

materials

MDPI

Article

The Effects of Spark-Plasma Sintering (SPS) on the Microstructure and Mechanical Properties of BaTiO$_3$/3Y-TZP Composites

Jing Li [1], Bencang Cui [1], Huining Wang [2], Yuanhua Lin [1,*], Xuliang Deng [3], Ming Li [1] and Cewen Nan [1]

[1] State Key Laboratory of New Ceramics and Fine Processing, School of Materials Science and Engineering, Tsinghua University, Beijing 100084, China; ljing12@mails.tsinghua.edu.cn (J.L.); cbc14@mails.tsinghua.edu.cn (B.C.); lim@mail.tsinghua.edu.cn (M.L.); cwnan@mail.tsinghua.edu.cn (C.N.)

[2] Department of Periodontics, Hospital of Stomatology Wenzhou Medical University, Wenzhou 325027, China; wanghuining1973@gmail.com

[3] Department of Geriatric Dentistry, School & Hospital of Stomatology, Peking University, Beijing 100081, China; kqdengxuliang@bjmu.edu.cn

* Correspondence: linyh@mail.tsinghua.edu.cn; Tel.: +86-10-6277-1160

Academic Editor: Eugene A. Olevsky

Received: 5 April 2016; Accepted: 25 April 2016; Published: 28 April 2016

Abstract: Composite ceramics BaTiO$_3$/3Y-TZP containing 0 mol %, 3 mol %, 5 mol %, 7 mol %, and 10 mol % BaTiO$_3$ have been prepared by conventional sintering and spark-plasma sintering (SPS), respectively. Analysis of the XRD patterns and Raman spectra reveal that the phase composition of t-ZrO$_2$, m-ZrO$_2$, and BaTiO$_3$ has been obtained. Our results indicate that SPS can be effective for the decrease in grain size and porosity compared with conventional sintering, which results in a lower concentration of m-ZrO$_2$ and residual stress. Therefore, the fracture toughness is enhanced by the BaTiO$_3$ phase through the SPS technique, while the behavior was impaired by the piezoelectric second phase through conventional sintering.

Keywords: spark-plasma sintering (SPS); BaTiO$_3$/3Y-TZP; fracture toughness

1. Introduction

As a field-assisted sintering technique, spark-plasma sintering (SPS) has attracted much attention since its advent in the late 1970s [1–3]. The starting powders in graphite die are sintered directly instead of being pre-pressed prior to sintering using the conventional processing technique. After graphite die is placed in the furnace, two pistons acting as electrodes load pressure on the upper and bottom surfaces. Due to the good electrical and thermal conductivity of the graphite die, adequate Joule heat is efficiently and quickly transferred to the starting powder under a relatively low voltages. Moreover, the heating rate can be as high as 1000 °C/min, resulting from the adjustable current pulses (milliseconds) [1–3].

Both the compressive press and high heating rate work to obtain dense bulks with nano-size grains under a lower sintering temperature, leading to the extensive application of SPS in the dielectric, piezoelectric, and thermoelectric fields, among others [4–8]. Li *et al.* [9] prepared lead-free piezoelectric ceramics Na$_{0.5}$K$_{0.5}$NbO$_3$ with 99% relative density at 920 °C by using the SPS technique. It was fairly difficult to obtain when the ceramic sintered in a conventional furnace. Deng *et al.* [10] synthesized nanostructured bulk BaTiO$_3$ with a grain size of 20 nm and a relative density of 97% via SPS, while the grain size of BaTiO$_3$ was in micrometer range when sintered in a conventional furnace. Furthermore, a series of transparent ceramics [11], such as alumina [12], zirconia [13], and yttrium-aluminum-garnet [14], have been processed with aid of the SPS technique.

Fracture toughness is a highly essential behavior for dental materials [15]. It is believed that the piezoelectric addition could enhance the toughness of ceramics based on the piezoelectric secondary phase toughening mechanism [16–18]. Under load, the piezoelectric effect would lead to domain wall motion and dissipate energy in the tips of cracks. Yang *et al.* [19] studied the effect of the piezoelectric second phase, $Nd_2Ti_2O_7$, on the fracture toughness of Al_2O_3, and the toughness increased to 6.7 MPa·$m^{1/2}$. Chen *et al.* [20] prepared a $Sr_2Nb_2O_7$/3Y-TZP composite and found that the fracture toughness was significant higher than that of 3Y-TZP, as high as 13 MPa·$m^{1/2}$. However, Yang *et al.* [21] found that the addition of $BaTiO_3$ suppressed the effect of the transformation toughening of 3Y-TZP, and the fracture toughness decreased instead of increased. Moreover, it has been proven that the electrical charges have a positive effect on the growth and differentiation of osteoblast cells, resulting from preferential adsorption of ions and proteins onto the polarized surfaces [22,23]. As piezoelectric materials could vary surface charges under load, the piezoelectric addition might induce improved bone formation around restorations.

The aim of this study was to investigate effects of the SPS technique on the microstructure, and mechanical properties of $BaTiO_3$/3Y-TZP composites as a function of $BaTiO_3$ content. The null hypotheses of this study were that SPS would help synthesize dense $BaTiO_3$/3Y-TZP bulks with nano-size grains and would improve the mechanical behaviors.

2. Results and Discussion

2.1. Phase Structure Analysis

The X-ray diffraction (XRD) patterns of $BaTiO_3$/3Y-TZP specimens prepared by different sintering techniques are shown in Figure 1. In this study, CS stands for a conventionally sintered specimen that is sintered in a conventional air furnace. All of the XRD patterns present a crystalline phase of tetragonal ZrO_2, and the characteristic peaks can be indexed to PDF card #50-1089. With respect to specimens sintered via the SPS technique, diffraction peaks due to $BaTiO_3$ were detected as the contents of $BaTiO_3$ increased to 7 mol % and 10 mol %. No m-ZrO_2 (monoclinic ZrO_2) phase was observed. However, as far as conventionally sintered specimens are concerned, peaks can be attributed to $BaTiO_3$, and m-ZrO_2 was detected with increasing content of $BaTiO_3$. Peaks at 39° can be indexed to the m-ZrO_2 phase and $BaTiO_3$ phase. According to the relative intensity of peaks of each phase, it is reasonable to find that the intensity of a peak at 39° mainly originates from the m-ZrO_2 phase. Content of the m-ZrO_2 phase was significantly lower in the spark-plasma-sintered samples; therefore, peaks at 39° seem to be absent in the spark-plasma-sintered samples. Moreover, the dominant ZrO_2 structure seems to be a monoclinic phase rather than a tetragonal phase when the contents of $BaTiO_3$ increase to 7 mol % and 10 mol %, demonstrating a quite different phase structure.

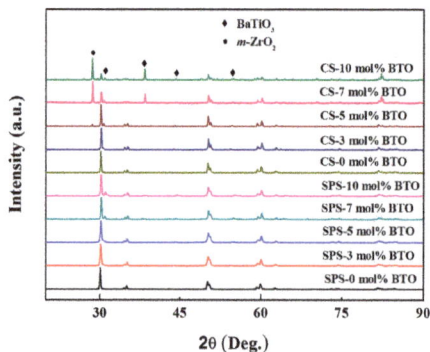

Figure 1. X-ray diffraction (XRD) patterns of spark-plasma-sintered and conventionally sintered $BaTiO_3$/3Y-TZP ceramics.

Furthermore, the T-M phase transformation within zirconia is investigated by one of the most effective techniques: Raman spectroscopy [24]. Figure 2a shows the Raman spectrum of a representative specimen. Characteristic peaks at wavenumbers 147 cm^{-1}, 265 cm^{-1}, 464 cm^{-1}, and 642 cm^{-1} represent t-ZrO$_2$, and peaks at wavenumbers 181 cm^{-1} and 190 cm^{-1} reveal the presence of m-ZrO$_2$ [25]. Figure 2b shows a volume fraction of m-ZrO$_2$ calculated according to Tabares and Anglada [26]. The content of m-ZrO$_2$ increases with BaTiO$_3$ concentration, which is consistent with the XRD patterns. Compared with conventionally sintered ceramics, spark-plasma-sintered specimens have much lower concentrations of m-ZrO$_2$, ranging between 4% and 29.4%, whereas the m-ZrO$_2$ content of conventionally sintered composites varies from 8% to 71.2%. The great discrepancy of the phase structure between specimens may originate from different residual stress states caused by the addition of BaTiO$_3$. Less m-ZrO$_2$ content is better for zirconia ceramics based on the well-known phase transformation toughening mechanism [27]; therefore, the low concentration of m-ZrO$_2$ might have a positive effect on the fracture toughness of spark-plasma-sintered composites.

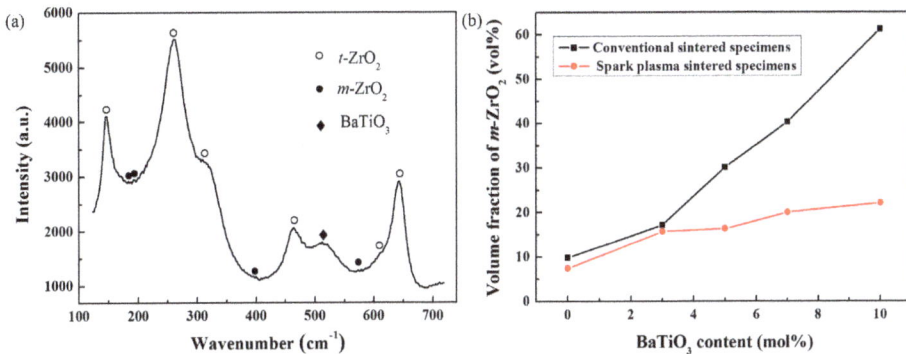

Figure 2. (a) Raman spectrum of BaTiO$_3$/3Y-TZP (with 7 mol % BaTiO$_3$) prepared via the SPS method; (b) Volume fraction of m-ZrO$_2$ content of spark-plasma-sintered and conventionally sintered BaTiO$_3$/3Y-TZP ceramics.

2.2. Microstructure Analysis

Figure 3 shows scanning election microscopy (SEM) images of spark-plasma-sintered and conventionally sintered BaTiO$_3$/3Y-TZP ceramics as a function of BaTiO$_3$ content. The As for conventionally sintered composites, two different kinds of grains, ZrO$_2$ and BaTiO$_3$ are clearly observed. With increasing amounts of BaTiO$_3$, the grain size of ZrO$_2$ is about 200 nm, remaining substantially unchanged, while grain size of BaTiO$_3$ increases from 1 to 3.5 µm with the BaTiO$_3$ content. Since grain size of BaTiO$_3$ is 5 to 17.5 times larger than that of ZrO$_2$, the mismatch of grain size leads to pores in ceramics and might introduce stress between grains, resulting in the stress-induced T-M phase transformation. Hence, the content of m-ZrO$_2$ increases, as shown in Figure 2. Compared with conventionally sintered composites, spark-plasma-sintered ceramics exhibit significantly smaller grain sizes, especially for BaTiO$_3$ grains, which can be attributed to the compressive press and the high heating rate. Moreover, with close ion radii and the same valence, it is likely for Ti^{4+} to partially substitute Zr^{4+}, forming a solid solution, (Zr,Ti)O$_2$ [21]. Thus, spark-plasma-sintered ceramics show traces of liquid-phase sintering with increasing content of BaTiO$_3$, resulting in more dense composite bulks [28]. The larger grain size for higher BaTiO$_3$ content in conventionally sintered specimens may also support this conclusion.

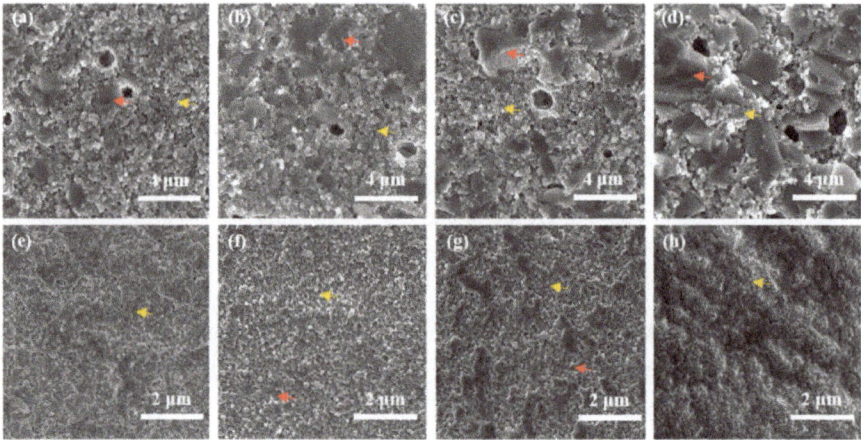

Figure 3. Scanning election microscopy (SEM) images of fracture surfaces of spark-plasma-sintered and conventionally sintered BaTiO$_3$/3Y-TZP ceramics. (**a–d**) Specimens are prepared by conventional sintering method; (**e–h**) Specimens are prepared via the SPS method. (**a,e**) 3 mol % BaTiO$_3$; (**b,f**) 5 mol % BaTiO$_3$; (**c,g**) 7 mol % BaTiO$_3$; (**d,h**) 10 mol % BaTiO$_3$. The red arrow denotes the BaTiO$_3$ phase and the yellow arrow denotes the 3Y-TZP phase.

According to the relative density, bulk porosities of composites (Table 1) are calculated by using the following equation: $P = (1 - \rho) \times 100\%$, where P is the bulk porosity, and ρ is the relative density. With increasing amounts of BaTiO$_3$, porosity trends between conventionally sintered specimens, and spark-plasma-sintered composites vary greatly. Porosity increases with BaTiO$_3$ content, due to the mismatch of grain size in conventionally sintered ceramics. By contrast, both of the slight mismatches in grain size and the liquid-phase sintering produce effects on the spark-plasma-sintered composites; therefore, porosity values float slightly. However, it is reasonable to find that spark-plasma-sintered composites are denser than the conventionally sintered ones, which are in consistency with the observed SEM images.

Table 1. The porosity of spark-plasma-sintered and conventionally sintered BaTiO$_3$/3Y-TZP ceramics.

BaTiO$_3$ Content	0 mol %	3 mol %	5 mol %	7 mol %	10 mol %
Porosity (%) (CS)	2.7	5.9	9.6	12.5	14.8
Porosity (%) (SPS)	0.5	4.8	3.3	1.5	3.2

2.3. Mechanical Properties

The residual stress state of the composites could be effective for not only the phase transformation but also the fracture toughness. As an attempt to depict the residual stress state, Raman maps of each specimen (5 × 5 μm^2) were recorded, and the data were analyzed by MATLAB software. Specimens without BaTiO$_3$ serve as the control group, and the mean wavenumber is 145.5 cm^{-1}. A shift of the peak toward higher wavelength number indicates the presence of residual compressive stress, which helps crack closure. By contrast, a peak shift toward lower wavelength number reveals the presence of residual tensile stress. Moreover, a larger peak shift means a higher residual stress. As seen in Figure 4, a mixture of tensile and compressive stress is recorded in spark-plasma-sintered composites, but significantly more tensile stress is found in the conventionally sintered specimens. Moreover, the peak shift is larger for conventionally sintered specimens with more BaTiO$_3$ content.

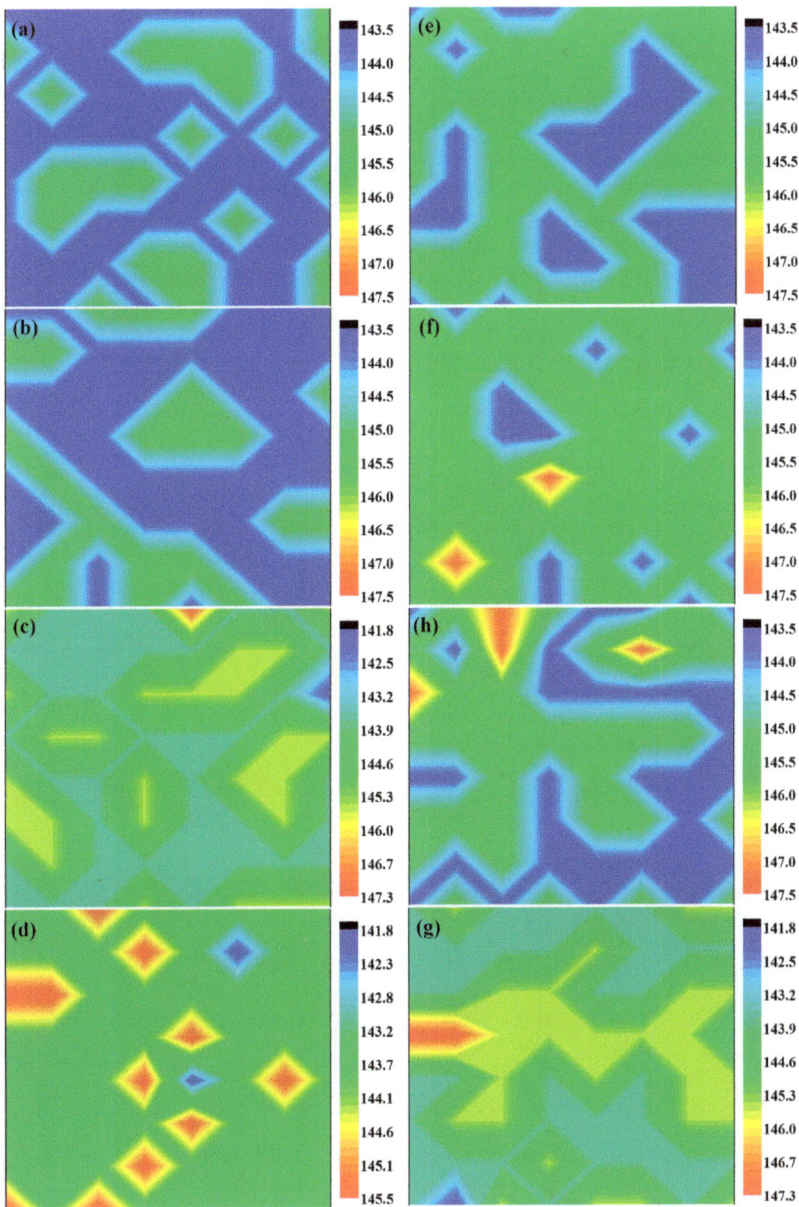

Figure 4. Quantitative Raman maps of spark-plasma-sintered and conventionally sintered $BaTiO_3$/3Y-TZP ceramics. (**a–d**) Specimens are prepared by conventional sintering method; (**e–h**) Specimens are prepared via the SPS method. (**a,e**) 3 mol % $BaTiO_3$; (**b,f**) 5 mol % $BaTiO_3$; (**c,g**) 7 mol % $BaTiO_3$; (**d,h**) 10 mol % $BaTiO_3$.

Figure 5 shows the fracture toughness, Vickers hardness, and elastic modulus values as a function of $BaTiO_3$ content. Even though the addition of $BaTiO_3$ would enhance the fracture toughness through the piezoelectric effect, both the accompanied high porosity and the m-ZrO_2 content have negative effects on the behavior of specimens prepared by conventional sintering. Therefore, the fracture toughness greatly trends downward after increasing slightly. Owing to the compressive press and the high heating rate of the SPS technique, the grain size of BaTiO$_3$ is quite smaller, and ceramics show traces of liquid-phase sintering, resulting in more dense composites and a low concentration of m-ZrO$_2$. Thus, the expected coupling effects of the piezoelectric secondary phase toughening mechanism and phase transformation toughening mechanism lead to a high toughness of the specimens prepared via the SPS method. As for composites with 3 mol % BaTiO$_3$, the fracture toughness of the conventionally sintered specimen is significantly higher than that of the spark-plasma-sintered specimen, which may be attributed to the difference in the effect of the piezoelectric secondary phase toughening mechanism, resulting from different contents of the BaTiO$_3$ phase. XRD patterns of both specimens show no BaTiO$_3$ phase, but SEM images reveal the existence of BaTiO$_3$ grains in the conventionally sintered specimen (Figure 3a). The addition of 3 mol % BaTiO$_3$ in the spark-plasma-sintered specimen may serve as a doping agent rather than a polycrystalline phase, which destroys the fracture toughness.

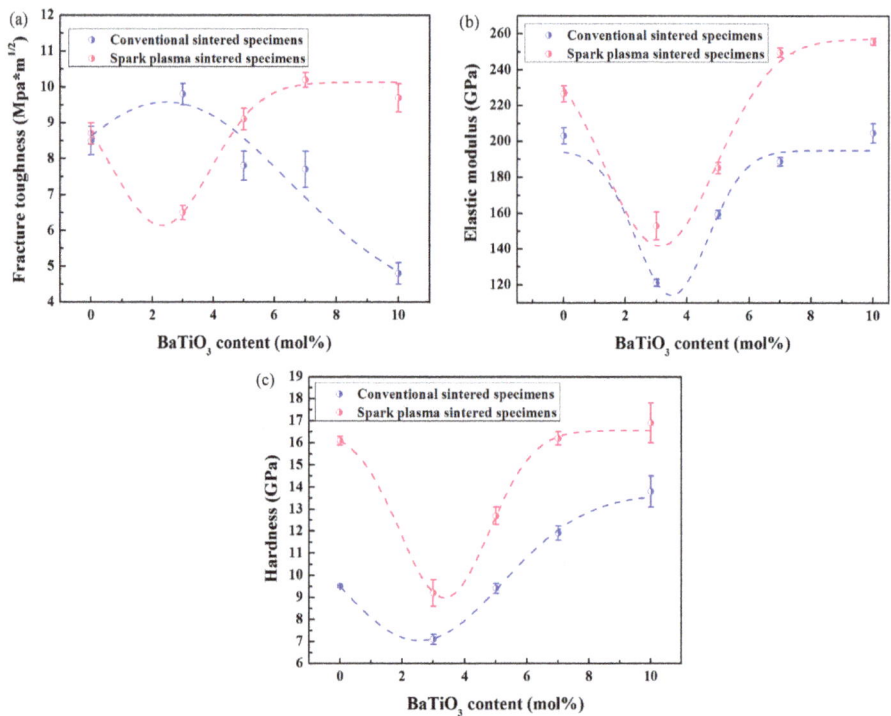

Figure 5. The mechanical properties of specimens with different BaTiO$_3$ contents. (a) Fracture toughness; (b) elastic modulus; (c) hardness.

Both the elastic modulus and hardness show a similar trend with increasing amounts of BaTiO$_3$. The as-prepared BaTiO$_3$ has a lower elastic modulus (~72.8 GPa) and hardness (~1.4 GPa) than does 3Y-TZP, but there is no expected decline in the composites. After decreasing at first, these two behaviors increase with BaTiO$_3$ content, which may be attributed to the formation of the solid solution, (Zr,Ti)O$_2$. As spark-plasma-sintered ceramics suffer a high compressive press during sintering, they show more traces of liquid-phase sintering, suggesting more amounts of solid solution [29]. Therefore, these specimens reveal a higher elastic modulus and hardness, compared with the conventionally sintered specimens.

3. Materials and Methods

3.1. Materials

3Y-TZP (TZ-3YSB-E, Tosoh Co., Tokyo, Japan) with an average particle size of 90 nm and BaTiO$_3$ (Sinopharm Chemical Reagent Co., Shanghai, China) with an average particle size of 100 nm were used to prepare the BaTiO$_3$/3Y-TZP composite.

3.2. Preparation of Porous Zirconia Ceramic

The starting materials, 3Y-TZP and BaTiO$_3$, at 0 mol %, 3 mol %, 5 mol %, 7 mol %, and 10 mol % were mixed together by alcohol-based ball milling for 12 h, respectively. The mixture powders were dried in oven for sintering. Some of the mixtures were pressed at a pressure of 4 MPa, followed by a cold isostatic pressing at 200 MPa, and the samples were then heated up to 1400 °C at a rate of 100 °C/h and kept for 2 h in a conventional air furnace. Some of the mixtures were sintered directly via SPS. The heating rate was 110 °C/s, and the sintering temperature was 1175 °C.

3.3. Characterization

X-ray diffraction spectroscopy (Rigaku, D/MAX-2550V, Tokyo, Japan) was employed to analyze the phase composition. Morphologies of fracture surfaces were examined via SEM (Hitachi, S-2500N, Tokyo, Japan). The relative density was measured by the Archimedes method. The volume fraction of monoclinic ZrO$_2$ was measured by a Raman spectrometer ((Hiroba, LabRAM HR Evolution, Tokyo, Japan). It was calculated based on the equation:

$$V_m = \frac{I_m^{181} + I_m^{190}}{0.32\left(I_t^{147} + I_t^{265}\right) + I_m^{181} + I_m^{190}} \tag{1}$$

where I_t and I_m are the integrated intensities of the tetragonal and monoclinic peaks, respectively. A nano-indentation tester (MTS, Palo Alto, CA, USA) was applied to analysis of the Vickers hardness and elastic modulus. The hardness was calculated by the equation:

$$H_V = 1.8544 P/d^2 \tag{2}$$

where H_V is the Vickers hardness; P is the load; and d is the diagonal of the indentation. The elastic modulus was further inferred by using the equation:

$$E = 0.45 H_V/(a/b - a/b_1) \tag{3}$$

where E is the elastic modulus; b is the length of the shorter diagonal; b_1 is the length of the longer diagonal; and a is the length of the crack. A universal test instrument (Shimadzu, EZ-100, Tokyo, Japan) was employed to measure the fracture toughness of specimens by the single-edge-notched beam method with a loading rate of 0.05 mm/min. Bending bars ($n = 12$) per specimen were cut into $2 \times 4 \times 16$ mm^3 with a diamond blade, and the notch depth was approximately 2 mm. The fracture toughness was calculated using formula:

$$K_{IC} = \frac{P_0 \times l}{BW^{3/2}} f\left(\frac{a}{W}\right) \tag{4}$$

where P_0 is the load; l is the span; B is the height of bar; W is the width of the bar; and a is the depth of the notch.

4. Conclusions

A series of BaTiO$_3$/3Y-TZP ceramics have been prepared by conventional sintering and SPS, respectively. The phase structure, microstructure, and mechanical properties of the composites were investigated as a function of BaTiO$_3$ content. Our results show that the SPS technique has a remarkably positive effect on the behaviors of BaTiO$_3$/3Y-TZP composites. Spark-plasma-sintered specimens are superior in fracture toughness due to the coupling effects of the piezoelectric secondary phase toughening mechanism and the phase transformation toughening mechanism. These results reveal that the piezoelectric secondary phase, BaTiO$_3$, could enhance the fracture toughness of zirconia through the SPS technique.

Acknowledgments: This work was a result of collaboration between the Materials Science and Engineering School in Tsinghua University, Department of Periodontics in Hospital of Stomatology Wenzhou Medical University and the Outpatient Dental Center in Peking University. It was supported by the National Science and Technology Ministry of China (Grant no. 2012BAI07B00).

Author Contributions: Huining Wang, Yuanhua Lin, Xuliang Deng, Ming Li, and Cewen Nan organized the research; Jing Li and Bencang Cui performed the experiments; Jing Li wrote the manuscript; all authors reviewed the manuscript.

Conflicts of Interest: The authors declare no conflict of interest.

References

1. Hulbert, D.M.; Anders, A.; Andersson, J.; Lavernia, E.J.; Mukherjee, A.K. A discussion on the absence of plasma in spark plasma sintering. *Scr. Mater.* **2009**, *60*, 835–838. [CrossRef]
2. Orrù, R.; Licheri, R.; Locci, A.M.; Cincotti, A.; Cao, G. Consolidation/synthesis of materials by electric current activated/assisted sintering. *Mater. Sci. Eng. R-Rpt.* **2009**, *63*, 127–287. [CrossRef]
3. Hulbert, D.M.; Anders, A.; Dudina, D.V.; Andersson, J.; Jiang, D.; Unuvar, C.; Anselmi-Tamburini, U.; Lavernia, E.J.; Mukherjee, A.K. The absence of plasma in "spark plasma sintering". *J. Appl. Phys.* **2008**, *104*, 033305. [CrossRef]
4. Biswas, K.; He, J.; Blum, I.D.; Wu, C.I.; Hogan, T.P.; Seidman, D.N.; Dravid, V.P.; Kanatzidis, M.G. High-performance bulk thermoelectrics with all-scale hierarchical architectures. *Nature* **2012**, *489*, 414–418. [CrossRef] [PubMed]
5. Munir, Z.A.; Anselmi-Tamburini, U.; Ohyanagi, M. The effect of electric field and pressure on the synthesis and consolidation of materials: A review of the spark plasma sintering method. *J. Mater. Sci.* **2006**, *41*, 763–777. [CrossRef]
6. Zhou, M.; Li, J.F.; Kita, T. Nanostructured AgPb$_m$SbTe$_{m+2}$ system bulk materials with enhanced thermoelectric performance. *J. Am. Chem. Soc.* **2008**, *130*, 4527–4532. [CrossRef] [PubMed]
7. Zhang, B.-P.; Li, J.-F.; Wang, K.; Zhang, H. Compositional dependence of piezoelectric properties in Na$_x$k$_{1-x}$NbO$_3$ lead-free ceramics prepared by spark plasma sintering. *J. Am. Ceram. Soc.* **2006**, *89*, 1605–1609. [CrossRef]
8. Yanagiya, S.; Van Nong, N.; Xu, J.; Pryds, N. The effect of (Ag, Ni, Zn)-addition on the thermoelectric properties of copper aluminate. *Materials* **2010**, *3*, 318–328. [CrossRef]
9. Li, J.-F.; Wang, K.; Zhang, B.-P.; Zhang, L.-M. Ferroelectric and piezoelectric properties of fine-grained Na$_{0.5}$k$_{0.5}$NbO$_3$ lead-free piezoelectric ceramics prepared by spark plasma sintering. *J. Am. Ceram. Soc.* **2006**, *89*, 706–709. [CrossRef]
10. Deng, X.; Wang, X.; Wen, H.; Kang, A.; Gui, Z.; Li, L. Phase transitions in nanocrystalline barium titanate ceramics prepared by spark plasma sintering. *J. Am. Ceram. Soc.* **2006**, *89*, 1059–1064. [CrossRef]
11. Wang, S.F.; Zhang, J.; Luo, D.W.; Gu, F.; Tang, D.Y.; Dong, Z.L.; Tan, G.E.B.; Que, W.X.; Zhang, T.S.; Li, S.; *et al.* Transparent ceramics: Processing, materials and applications. *Prog. Solid State Chem.* **2013**, *41*, 20–54. [CrossRef]
12. Kim, B.N.; Hiraga, K.; Morita, K.; Yoshida, H. Spark plasma sintering of transparent alumina. *Scr. Mater.* **2007**, *57*, 607–610. [CrossRef]
13. Alaniz, J.E.; Perez-Gutierrez, F.G.; Aguilar, G.; Garay, J.E. Optical properties of transparent nanocrystalline yttria stabilized zirconia. *Opt. Mater.* **2009**, *32*, 62–68. [CrossRef]

14. Chaim, R.; Kalina, M.; Shen, J.Z. Transparent yttrium aluminum garnet (yag) ceramics by spark plasma sintering. *J. Eur. Ceram. Soc.* **2007**, *27*, 3331–3337. [CrossRef]

15. Denry, I.; Holloway, J. Ceramics for dental applications: A review. *Materials* **2010**, *3*, 351–368. [CrossRef]

16. Chen, X.M.; Yang, B. A new approach for toughening of ceramics. *Mater. Lett.* **1997**, *33*, 237–240. [CrossRef]

17. Liu, Y.G.; Jia, D.C.; Zhou, Y. Microstructure and mechanical properties of a lithium tantalate-dispersed -alumina ceramic composite. *Ceram. Int.* **2002**, *28*, 111–114. [CrossRef]

18. Li, J.Y.; Dai, H.; Zhong, X.H.; Zhang, Y.F.; Ma, X.F.; Meng, J.; Cao, X.Q. Lanthanum zirconate ceramic toughened by BaTiO$_3$ secondary phase. *J. Alloy. Compd.* **2008**, *452*, 406–409. [CrossRef]

19. Yang, B.; Chen, X.M. Alumina ceramics toughened by a piezoelectric secondary phase. *J. Eur. Ceram. Soc.* **2000**, *20*, 1687–1690. [CrossRef]

20. Liu, X.Q.; Chen, X.M. Microstructures and mechanical properties of Sr$_2$Nb$_2$O$_7$-toughened 3Y-TZP ceramics. *Ceram. Int.* **2003**, *29*, 635–640. [CrossRef]

21. Yang, B.; Chen, X.M.; Liu, X.Q. Effect of BaTiO$_3$ addition on structures and mechanical properties of 3Y-TZP ceramics. *J. Eur. Ceram. Soc.* **2000**, *20*, 1153–1158. [CrossRef]

22. Hastings, G.W.; Mahmud, F.A. The electromechanical properties of fluid-filled bone: A new dimension. *J. Mater. Sci. Mater. Med.* **1991**, *2*, 118–124. [CrossRef]

23. Feng, J.Q.; Yuan, H.P.; Zhang, X.D. Promotion of osteogenesis by a piezoelectric biological ceramic. *Biomaterials* **1997**, *18*, 1531–1534. [CrossRef]

24. Li, M.; Feng, Z.; Xiong, G.; Ying, P.; Xin, Q.; Li, C. Phase transformation in the surface region of zirconia detected by UV Raman spectroscopy. *J. Phys. Chem. B* **2001**, *105*, 8107–8111. [CrossRef]

25. Djurado, E.; Bouvier, P.; Lucazeau, G. Crystallite size effect on the tetragonal-monoclinic transition of undoped nanocrystalline zirconia studied by XRD and Raman spectrometry. *J. Solid State Chem.* **2000**, *149*, 399–407. [CrossRef]

26. Muñoz Tabares, J.A.; Anglada, M.J. Quantitative analysis of monoclinic phase in 3Y-TZP by Raman spectroscopy. *J. Am. Ceram. Soc.* **2010**, *93*, 1790–1795. [CrossRef]

27. Garvie, R.C.; Hannink, R.H.; Pascoe, R.T. Ceramic steel? *Nature* **1975**, *258*, 703–704. [CrossRef]

28. Maneshian, M.H.; Simchi, A. Solid state and liquid phase sintering of mechanically activated W–20 wt. % Cu powder mixture. *J. Alloy. Compd.* **2008**, *463*, 153–159. [CrossRef]

29. Sivaprahasam, D.; Chandrasekar, S.B.; Sundaresan, R. Microstructure and mechanical properties of nanocrystalline WC–12Co consolidated by spark plasma sintering. *Int. J. Refract. Met. Hard Mater.* **2007**, *25*, 144–152. [CrossRef]

![materials logo] *materials*

MDPI

Article

Liquid Film Capillary Mechanism for Densification of Ceramic Powders during Flash Sintering

Rachman Chaim

Department of Materials Science and Engineering, Technion-Israel Institute of Technology, Haifa 32000, Israel; rchaim@technion.ac.il; Tel.: +972-4-829-4589

Academic Editor: Eugene A. Olevsky
Received: 21 March 2016; Accepted: 7 April 2016; Published: 11 April 2016

Abstract: Recently, local melting of the particle surfaces confirmed the formation of spark and plasma during spark plasma sintering, which explains the rapid densification mechanism via liquid. A model for rapid densification of flash sintered ceramics by liquid film capillary was presented, where liquid film forms by local melting at the particle contacts, due to Joule heating followed by thermal runaway. Local densification is by particle rearrangement led by spreading of the liquid, due to local attractive capillary forces. Electrowetting may assist this process. The asymmetric nature of the powder compact represents an invasive percolating system.

Keywords: flash sintering; spark plasma sintering; densification; melting; electric field; electric conductivity; ceramics; invasion percolation

1. Introduction

Spark plasma sintering (SPS) and Flash sintering (FS) are nowadays considered important processes for rapid densification of ceramic particles to fully dense solids. The two techniques differ by the set-up, the ranges of the applied voltage and the electric current density; however, the physical and chemical processes, and the electric field effects with the ionic/semiconducting ceramics are very similar. In this respect, different densification mechanisms suggested for the enhanced densification kinetics, where the processes at the particle surfaces and interfaces play the dominating role [1–12]. Recently, local melting of ceramic particle surfaces further supported the formation of spark and plasma during SPS [13–17]. Therefore, spark and plasma related to discharge of surface charges, accumulated on the non-conducting particle surfaces [18]. Consequently, no spark and plasma were expected in conducting ceramic particles subjected to SPS; the enhanced densification in the later systems is associated with excessive Joule heating at the particle contact points. This latter behavior, where a contiguous particle network provides a percolative path for electric conduction, is tangent to the FS process. Thus, Joule heating followed by thermal runaway are the main current explanations for flash sintering. Yet, densification in FS specimens lacks the corresponding rapid densification mechanism.

Although extensive efforts have been invested towards characterization of the FS parameters and the voltage/current behavior, less attention has been paid to understanding of the involved densification mechanisms, with respect to the observed microstructure. In this regard, three major problems exist: first, most of the analyses relate the actual local process temperature to the temperature of the specimen, either technically measured (thermocouple, pyrometer), or calculated from the average value of the current (*i.e.*, black body calculation from electrical energy dissipation), or indirect measurements (*i.e.*, thermal expansion calibrations [19]). Consequently, the actual process temperature at the particle contacts is uncertain, and its estimation may vary by hundreds of degrees from work to work [8,19–21]. Second is the assumption that sintering takes place at once throughout the specimen. This assumption originates from the lack of interrupted FS experiments, or incorrect interpretation of the resulting microstructures. The third is the assumption about solid state sintering during the FS process. Simple calculation of the diffusion distances necessary for mass transport, using the ionic diffusion coefficients fail to explain the densification within the typical time intervals of flash sintering by solid state sintering.

The yield stress and electrical conductivity, and their temperature dependence, are the main properties that determine the conditions of the field effects, *i.e.*, at what conditions spark and plasma will form in a given non-conducting granular compact. These properties are used to construct plastic-deformation plasma-formation temperature window diagrams [13,14,18,22], from which one can deduce the SPS pressure-temperature schedule appropriate for enhanced densification. This introduces a new region of enhanced densification by plasma, with a transient nature, into the conventional densification mechanism maps. The relative location of this new region is determined by the plasma-formation temperature window; its position is expected to shift due to the particle size dependence of the yield strength and the surface conductivity, as well as the value of the applied pressure. Therefore, this transient state of rapid densification in SPS was previously termed "surface softening", due to the uncertainty concerning the nature of the liquid at the particle surfaces [23].

Following several reported microstructures of ceramics densified via FS, one can observe crystal growth from remnant of liquid phase [24], crystal growth from vapor [25], and curved boundaries, which are characteristic of a liquid presence during sintering [26]. Furthermore, the local nature of melting/reaction was confirmed by the islands in the partially reacted microstructure [9]. Several works also reported the formation of local liquid as well as heterogeneous microstructure during flash sintering [9,25]. Therefore, in this paper I will evaluate the formation of liquid at the particle contacts during FS and its consequences upon densification kinetics and microstructure via capillary forces.

2. Analysis and Discussion

2.1. Spark Plasma Sintering

The formation of spark and plasma depends on many material and process parameters, and takes place as a transient phenomenon during the densification. The charged particle surfaces at the compact cavities act as sources for spark and plasma [18], when they attain certain conditions for percolation of the electric current. The ignited plasma enhances densification under minimal applied pressure, via particles sliding aided by a "softened layer" (here will be treated as a liquid film) at their surfaces. The sudden increase in density and particle connectivity suppresses further expansion of the plasma; hence, it represents a local process with transient non-equilibrium character. Therefore, plasma region in the deformation mechanism map is a sort of "ladder" for climbing to higher densities within extremely short durations.

When spark and plasma increase the surface temperature so that a liquid film forms at the particle surfaces, the particles can sinter together in the absence of external pressure, or slide over each other by viscous flow, in the presence of an external pressure [13]. The resistance to mutual sliding of the two solid particles covered with a liquid film depends on the relative thickness of the liquid film compared to the particles radius [27]. Jagota and Dawson [27] treated this problem and showed that under certain conditions (*i.e.*, liquid film thickness to particle radius ratio higher than 0.2) the system behaves if no solid particles exist, *i.e.*, densification may take place by simple viscous sintering. However, such liquid layer thicknesses are unseen, neither in SPS, nor in FS, and I restrict the discussion to melting of a few atomic layers (*i.e.*, film) at the contact points between particles.

2.2. Flash Sintering

The specimen set-up in FS technique assumes compacted green powder through which the electric current passes via external electrodes, while the specimen is heated within the furnace. The green strength assures particle contiguity and available path for percolation of the electric current at the flash temperature. The main event of flash is a non-linear increase of the current at the onset flash temperature. Du *et al.* [10] attempt to explain this non-linearity as an artifact due to the estimation method of the specimen temperature. However, their measurements also refer to the average temperature instead of a local temperature in the specimen. Such non-linearity in the local current at the particle contacts may exist and explain the rapid sintering and densification, as shown below.

2.2.1. Local Capillary Forces

Recent FS works support the thermal runaway model as a dominating process at the flash set point [8,10,28]. Several significant processes may take place if the actual local temperature at the contact point surpasses the melting point of the solid particle at contact due to the Joule heating. First, the similar composition of the melt and the solid particle from which it melted leads to very low solid-liquid interfacial energy, hence full wetting of the solid particles by the melt at the contact point. The capillary forces associated with such a liquid layer depend on the solid particle size. For 100 nm Al_2O_3 nano-particles (in diameter) with liquid-vapor surface tension of 665 mJ·m^{-2} in air [29], attractive capillary force of ~27 MPa is estimated. This capillary force is of the same order of the pressures applied during the SPS and hot pressing, and is high enough to attract the adjacent particles and lead to their local rearrangement and compaction.

Another aspect of wetting which is worth noting is the occasions when local melting and wetting cause the formation of a gap (previously a contact) between the particles. In such a case, the presence of high local electric fields over the micrometric or nanometric gaps can lead to electrowetting [30] of the ionic melt, and affect both the liquid dihedral angle as well as the spreading degree of the liquid on the particle surfaces.

2.2.2. Local Electric Conductivity

Let us assume that the green compact of ceramic powder is heated and subjected to increasing current density at constant voltage. Such a system is composed of different resistances (particles and contact points) and capacitances (particle gaps). Once an appropriate electrical conductivity is gained, the higher electric resistivity at the particle contacts preferably consumes the current for the Joule heating. The local heating at the contact point increases the local temperature, and consequently the local electric conductivity (ionic and electronic), due to the negative temperature coefficient of resistivity. This process has an autocatalytic effect, which may eventually lead to local melting at the contact points. Formation of melt at the contact point also has significant implications on the FS process, which have been underestimated until now, if not neglected. The ionic conductivity in many oxides (*i.e.*, BeO, Al_2O_3, Sc_2O_3, ZrO_2, Y_2O_3) experiences an abrupt jump at melting, where the conductivity in the melt may increase by two to four orders of magnitude compared to the crystalline state [31,32]. The conductivity in pure oxide melts is controlled by cations, when ionic bonds dominate (*i.e.*, MgO, CaO, Li_2O, Al_2O_3), and by electrons, when significant covalent bonding exists, and leads to semiconducting behavior (*i.e.*, Bi_2O_3, CuO, MnO, TiO_2, V_2O_3). The electric conductivity (σ) in such melts depends on the bond strength, as well as on the size of the structural units moving within the melt, hence the melt viscosity (η). The product of the electric conductivity by the viscosity for a given ion is constant and expressed by Walden's rule [33]:

$$\sigma \cdot \eta = n_{i,melt} \frac{(Z \cdot e)^2}{3\pi d} = constant \tag{1}$$

where $n_{i,melt}$ is the number of the *i*-th ion per unit volume when the transport number is unity, Z is the ion charge, d is the ion diameter, and e is the unit charge of the electron. This model assumes that the conducting species are spherical ions that are in steady state flow. Nevertheless, viscosity models assume flow of structural units and associate species within the ceramic melts, rather than single ions. Using the viscosity of pure ceramic melts and their volume, one can estimate fairly well the electric conductivity at the corresponding temperature.

Here I will limit the discussion to the case of pure Al_2O_3, for which there is extensive reliable physical and chemical data in the literature and has been densified by flash sintering [34]. The structural units in alumina melt are $AlO_{1.5}$, which represent a typical oxygen octahedron around the Al^{+3} cation [35]. Assuming partial ionic conductivity, *i.e.*, transference number 0.5, via diffusion of aluminum octahedron in the melt, the equivalent ionic diameter using the octahedron volume (as a sphere) is 2.397 nm. These assumptions lead to the more conservative calculation of the electrical conductivity (*i.e.*, lower values). In addition, the number of the ions per unit volume in the melt decreases compared to that of the crystalline solid, due to the decrease in the melt density. Therefore, the number of ions per unit volume in Equation (1) normalized by:

$$n_{melt} = n_{solid} \frac{\rho_{melt}}{\rho_{soild}} \tag{2}$$

where ρ_i and n_i are the density, and the number of the aluminum ions per unit volume, respectively, in either the melt or the crystalline solid.

Combining Equations (1) and (2), and using the following data for alumina: n_{solid} = 6, ρ_{solid} = 3.98 g·cm^{-3}, d = 2.397 nm, e = 1.602 × 10^{-19} coul, the electric conductivity of alumina was calculated from its melt viscosity [35] and its melt volume change [36] *versus* temperature. These data, with some experimental conductivity values measured in air [37,38], are presented in Figure 1. The calculated electric conductivity values (blue dashed line) are higher only by one order of magnitude from the experimentally measured values of super pure Al_2O_3. Overall, melting of alumina was followed by increase of electric conductivity at the melt by at least two orders of magnitude. Therefore, local melting at the contact points may increase the local current density by two orders of magnitude. The probability for local melting depends on the relative local conductance of the contact points available within the green compact. The smaller the contact diameter, the higher the current density through it, hence the higher the probability for melting. This should also lead to lower onset temperature for the flash, consistent with the reported data on the particle size effects [39]. Consequently, local melting should first take place at the loci of smaller contacts (*i.e.*, smaller particles), within the green compact.

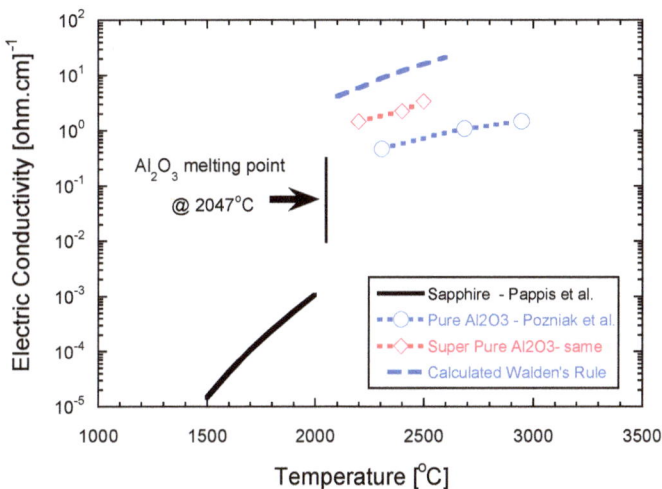

Figure 1. Electric conductivity of Sapphire crystal (black solid line) and alumina melt (blue dashed line) calculated from Walden's rule. The measured experimental values for pure (blue open circles) and super-pure alumina (red open diamonds) melts presented for comparison. Melting leads to increase in the electric conductivity by two to four orders of magnitude.

Since the liquid between the two contacting particles assumed to wet both particles, further melting should take place at the contacts of larger particles. Thus, the local melting progress is hierarchical, which assures the preservation of the local melt, as long as other percolative paths for the current flow exist. As was mentioned above, formation of a local melt leads to capillary forces high enough for local rearrangement of the surrounding particles. The local electrical resistance after melting, wetting, and rapid local densification is free of contact resistance, and falls to low values of the melt resistivity. Therefore, further dissipation of the electrical energy by the Joule heating will take place at other solid contacts having higher electrical resistances. This change in the local conductivity provides the conditions needed for promotion of this liquid assisted rearrangement and densification mechanism by capillary forces at different loci throughout the compact. The asymmetric nature of the powder compact subjected to electric current and connected to two different electrodes (*i.e.*, Cathode and Anode) represents an invasive percolative system [40]. In addition, the ceramic particles are most often characterized by a fractal character, which may change their electrical response. I will treat these aspects in a future paper since the present densification model of liquid film capillary is still valid.

2.2.3. Local Volume Change

Local melting also leads to significant increase in the specific volume, expressed by a decrease in the melt density, compared to that of the crystalline solid. The lattice parameters of Sapphire are c = 1.29915 nm, a = 0.47592 nm, using the hexagonal notation of the rhombohedral lattice. The thermal expansion coefficients of Sapphire above 1500 °C are almost constant, with values of $\alpha_c = 9 \times 10^{-6}$ °C^{-1} and $\alpha_a = 8 \times 10^{-6}$ °C^{-1} [41]. Using the lattice parameters of Sapphire and its thermal expansion coefficients, the changes in density (black solid line) and the specific volume (dashed blue line) were calculated and shown in Figure 2. The corresponding data for the melt density and its specific volume are also plotted in Figure 2, using data from the literature [36]. The specific volume in Al$_2$O$_3$ increases by ~20% at the melting temperature; further linear increase observed with temperature, albeit with much higher gradient than in the solid (dashed blue lines in Figure 2). Once melting takes place at the contact point, the local volume increase provides a liquid meniscus necessary for wetting of the adjacent solid particles, and aids their local rearrangement and densification.

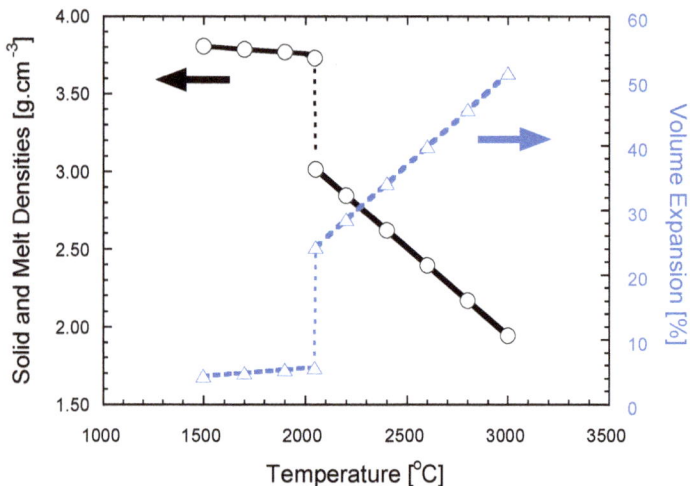

Figure 2. Densities of crystalline alumina and its melt (solid black line) *versus* temperature exhibits discontinuous decrease at the melting temperature. The corresponding increase in the specific volume expansion with temperature presented by the dashed blue line. Data for the melt density used from reference [36].

The discontinuous increase in both electrical conductivity and specific volume with melting are inherent physical properties of ceramic crystals. Therefore, all ceramic powder compacts subjected to an electric field are prone to flash sintering, once critical flash conditions attained. The flash onset condition is mainly controlled by the amount of the applied electric power density [28], and leads to local melting at the particle contacts. The non-linear electric conductivity is associated with this local melting. In this respect, simulations of flash sintering confirmed the thermal runaway to be a consequence of the temperature dependent resistivity [42]. Narayan proposed a model for grain boundary melting, albeit for description of grain growth during flash sintering [43]. Finally, striking similarities exist between the flash sintering kinetics of sub-micrometer pure alumina [44] and that of the same powder with 2 mol% liquid forming additives, subjected to conventional sintering [45]; this indicates the important role of the liquid phase during flash sintering.

3. Conclusions

A model of liquid-film capillary was introduced as a mechanism for the rapid densification during flash sintering. The thermal runaway due to the preferred Joule heating at the particle contacts leads to local melting at these loci, followed by particle wetting. The attractive capillary forces associated with this liquid film lead to particle rearrangement hence to densification. The rapid densification aided by the local increase both in the specific volume and in the electric conductivity due to the melt at the contacts. The contacts melt in a random hierarchical manner and the process has an asymmetric nature. The overall process is a critical phenomenon and modeled by an invasion percolation.

Conflicts of Interest: The author declares no conflict of interest.

Abbreviations

The following abbreviations used in this manuscript:

SPS   Spark Plasma Sintering
FS   Flash Sintering

References

1. Tokita, M. Mechanism of spark plasma sintering and its application to ceramics. *Nyn Seramikkasu* **1997**, *10*, 43–53.
2. Groza, J.R.; Zavaliangos, A. Sintering activation by external electrical field. *Mater. Sci. Eng. A* **2000**, *287*, 171–177. [CrossRef]
3. Munir, Z.A.; Anselmi-Tamburini, U.; Ohyanagi, M. The effect of electric field and pressure on the synthesis and consolidation of materials: A review of the spark plasma sintering method. *J. Mater. Sci.* **2006**, *41*, 763–777. [CrossRef]
4. Chaim, R. Densification mechanism in spark plasma sintering of nanocrystalline ceramics. *Mater. Sci. Eng. A* **2007**, *443*, 25–32. [CrossRef]
5. Olevsky, E.; Bogechev, I.; Maximenko, A. Spark-plasma sintering efficiency control by inter-particle contact area growth: A viewpoint. *Scr. Mater.* **2013**, *69*, 112–116. [CrossRef]
6. Holland, T.B.; Anselmi-Tamburini, U.; Mukherjee, A.K. Electric fields and the future of scalability in spark plasma sintering. *Scr. Mater.* **2013**, *69*, 117–121. [CrossRef]
7. Raj, R. Joule heating during flash sintering. *J. Eur. Ceram. Soc.* **2012**, *32*, 2293–2301. [CrossRef]
8. Todd, R.I.; Zapata-Solvas, E.; Bonilla, R.S.; Sneddon, T.; Wilshaw, P.R. Electrical characterization of flash sintering: Thermal runaway of Joule heating. *J. Eur. Ceram. Soc.* **2015**, *35*, 1865–1877. [CrossRef]
9. Batista Caliman, L.; Bouchet, R.; Gouvea, D.; Soudant, P.; Steil, M.C. Flash sintering of ionic conductors: The need of a reversible electrochemical reaction. *J. Eur. Ceram. Soc.* **2016**, *36*, 1253–1260. [CrossRef]
10. Du, Y.; Stevenson, A.J.; Vernat, D.; Diaz, M.; Marinha, D. Estimating Joule heating and ionic conductivity during flash sintering of 8YSZ. *J. Eur. Ceram. Soc.* **2016**, *36*, 749–759. [CrossRef]
11. Narayan, J. A new mechanism for field-assisted processing and flash sintering of materials. *Scr. Mater.* **2013**, *69*, 107–111. [CrossRef]

12. Naik, K.S.; Sglavo, V.M.; Raj, R. Flash sintering as a nucleation phenomenon and a model thereof. *J. Eur. Ceram. Soc.* **2014**, *34*, 4063–4067. [CrossRef]
13. Marder, R.; Estournès, C.; Chevallier, G.; Chaim, R. Plasma in spark plasma sintering of ceramic particle compacts. *Scr. Mater.* **2014**, *82*, 57–60. [CrossRef]
14. Marder, R.; Estournès, C.; Chevallier, G.; Chaim, R. Spark and plasma in spark plasma sintering of rigid ceramic nanoparticles: A model system of YAG. *J. Eur. Ceram. Soc.* **2015**, *35*, 211–218. [CrossRef]
15. Zhang, Z.; Liu, Z.; Lu, J.; Shen, X.; Wang, F.; Wang, Y. The sintering mechanism in spark plasma sintering—Proof of the occurrence of spark discharge. *Scr. Mater.* **2014**, *81*, 56–59. [CrossRef]
16. Saunders, T.; Grasso, S.; Reece, M.J. Plasma formation during electric discharge (50 V) through conductive powder compact. *J. Eur. Ceram. Soc.* **2015**, *35*, 871–877. [CrossRef]
17. Demirskyi, D.; Borodianska, H.; Grasso, S.; Sakka, Y.; Vasylkiv, O. Microstructure evolution during field-assisted sintering of zirconia spheres. *Scr. Mater.* **2011**, *65*, 683–686. [CrossRef]
18. Marder, R.; Estournès, C.; Chevallier, G.; Chaim, R. Numerical model for sparking and plasma formation during spark plasma sintering of ceramic compacts. *J. Mater. Sci.* **2015**, *50*, 4636–4645. [CrossRef]
19. Terauds, K.; Lebrun, J.M.; Lee, H.H.; Jeon, T.Y.; Lee, S.H.; Je, J.H.; Raj, R. Electroluminescence and the measurement of temperature during stage III of flash sintering experiments. *J. Eur. Ceram. Soc.* **2015**, *35*, 3195–3199. [CrossRef]
20. Grasso, S.; Sakka, Y.; Rendtorff, N.; Hu, C.; Maizza, G.; Borodianska, H.; Vasylkiv, O. Modeling of the temperature distribution of flash sintered zirconia. *J. Ceram. Soc. Jpn.* **2011**, *119*, 144–146. [CrossRef]
21. Muccillo, R.; Kleitz, M.; Muccillo, E.N.S. Flash grain welding in yttria-stabilized zirconia. *J. Eur. Ceram. Soc.* **2011**, *31*, 1517–1521. [CrossRef]
22. Chaim, R. Electric field effects during spark plasma sintering of ceramic nanoparticles. *J. Mater. Sci.* **2013**, *48*, 502–510. [CrossRef]
23. Chaim, R.; Marder-Jaeckel, R.; Shen, J.Z. Transparent YAG ceramics by surface softening of nanoparticles in spark plasma sintering. *Mater. Sci. Eng. A* **2006**, *429*, 74–78. [CrossRef]
24. Zhang, Y.; Jung, J.I.; Luo, J. Thermal runaway, flash sintering and asymmetrical microstructural development of ZnO and ZnO-Bi$_2$O$_3$ under direct currents. *Acta Mater.* **2015**, *94*, 87–100. [CrossRef]
25. Steil, M.C.; Marinha, D.; Aman, Y.; Gomes, J.R.C.; Kleitz, M. From conventional ac flash sintering of YSZ to hyper-flash and double flash. *J. Eur. Ceram. Soc.* **2013**, *33*, 2093–2101. [CrossRef]
26. Jha, S.K.; Lebrun, J.M.; Raj, R. Phase transformation in the alumina-titania system during flash sintering experiments. *J. Eur. Ceram. Soc.* **2016**, *36*, 733–739.
27. Jagota, A.; Dawson, P.R. Simulation of the viscous sintering of coated particles. *Acta Metall.* **1988**, *36*, 2551–2561. [CrossRef]
28. Bichaud, E.; Chaix, J.M.; Carry, C.; Kleitz, M.; Steil, M.C. Flash sintering incubation in Al$_2$O$_3$/TZP composites. *J. Eur. Ceram. Soc.* **2015**, *35*, 2587–2592. [CrossRef]
29. Lihrmann, J.M.; Haggerty, J.S. Surface tension of alumina-containing liquids. *J. Am. Ceram. Soc.* **1985**, *68*, 81–85. [CrossRef]
30. Mugele, F.; Baret, J.C. Electrowetting: From basics to applications. *J. Phys. Condens. Mater.* **2005**, *17*, R705. [CrossRef]
31. Lakomsky, V.I. Part 3: Oxide Cathodes for the Electric Arc. In *Welding and Surface Reviews*; Paton, B.E., Ed.; Harwood Academic Publishers: Reading, UK, 2000; Volume 13, pp. 74–90.
32. Leu, A.L.; Ma, S.M.; Eyring, H. Properties of molten magnesium oxide. *Proc. Natl. Acad. Sci. USA* **1975**, *72*, 1026–1030. [CrossRef] [PubMed]
33. Rennie, R. *Oxford Dictionary of Chemistry*, 6th ed.; Oxford University Press: Oxford, UK, 2016.
34. Biesuz, M.; Sglavo, V.M. Flash sintering of alumina: Effect of different operating conditions on densification. *J. Eur. Ceram. Soc.* **2016**. [CrossRef]
35. Wu, G.; Yazhenskikh, E.; Hack, K.; Wosch, E.; Müller, M. Viscosity model for oxide melts relevant to fuel slags. Part 1: Pure oxides and binary systems in the system SiO$_2$-Al$_2$O$_3$-CaO-MgO-Na$_2$O-K$_2$O. *Process. Tech.* **2015**, *137*, 93–103. [CrossRef]
36. Kirshenbaum, A.D.; Cahill, J.A. The density of liquid aluminum oxide. *J. Inorg. Nucl. Chem.* **1960**, *14*, 283–287. [CrossRef]
37. Pappis, J.; Kingery, W.D. Electrical properties of single-crystal and polycrystalline Alumina at high temperatures. *J. Am. Ceram. Soc.* **1961**, *44*, 459–464. [CrossRef]

38. Pozniak, I.; Pechenkov, A.; Shatunov, A. Electrical conductivity measurement of oxide melts. In Proceedings of the International Scientific Colloquium Modelling for Material Processing, Riga, Latvia, 8–9 June 2006.

39. Francis, J.S.C.; Cologna, M.; Raj, R. Particle size effects in flash sintering. *J. Eur. Ceram. Soc.* **2012**, *32*, 3129–3136. [CrossRef]

40. Wilkinson, D.; Willemsen, J.F. Invasion percolation: A new form of percolation theory. *J. Phys. A Math. Gen.* **1983**, *16*, 3365–3374. [CrossRef]

41. Dobrovinskaya, E.R.; Lytvynov, L.A.; Pishchik, V. *Sapphire: Material, Manufacturing, Applications*; Springer Science + Business Media: New York, NY, USA, 2009.

42. Da Silva, J.G.P.; Al-Qureshi, H.A.; Keil, F.; Janssen, R. A dynamic bifurcation criterion for thermal runaway during the flash sintering of ceramics. *J. Eur. Ceram. Soc.* **2016**, *36*, 1261–1267. [CrossRef]

43. Narayan, J. Grain growth model for electric field-assisted processing and flash sintering of materials. *Scr. Mater.* **2013**, *68*, 785–788. [CrossRef]

44. Gurt Santanach, J.; Weibel, A.; Estournès, C.; Yang, Q.; Laurent, C.; Peigney, A. Spark plasma sintering of alumina: Study of parameter, formal sintering analysis and hypotheses on the mechanism(s) involved in densification and grain growth. *Acta Mater.* **2011**, *59*, 1400–1408. [CrossRef]

45. Xue, L.A.; Chen, I.W. Low-temperature sintering of alumina with liquid-forming additives. *J. Am. Ceram. Soc.* **1991**, *74*, 2011–2013. [CrossRef]

materials

MDPI

Article

On the Mechanism of Microwave Flash Sintering of Ceramics

Yury V. Bykov [1], **Sergei V. Egorov** [1], **Anatoly G. Eremeev** [1], **Vladislav V. Kholoptsev** [1], **Ivan V. Plotnikov** [1], **Kirill I. Rybakov** [1,2,*] **and Andrei A. Sorokin** [1]

[1]   Institute of Applied Physics, Russian Academy of Sciences, Nizhny Novgorod 603950, Russia;
      byk@appl.sci-nnov.ru (Y.V.B.); egr@appl.sci-nnov.ru (S.V.E.); aeremeev@appl.sci-nnov.ru (A.G.E.);
      holo@appl.sci-nnov.ru (V.V.K.); ivanplotnikov@appl.sci-nnov.ru (I.V.P.); asorok@appl.sci-nnov.ru (A.A.S.)
[2]   Advanced School of General and Applied Physics, Lobachevsky State University of Nizhny Novgorod,
      Nizhny Novgorod 603950, Russia
*   Correspondence: rybakov@appl.sci-nnov.ru; Tel.: +7-831-416-4831

Academic Editor: Eugene A. Olevsky
Received: 12 May 2016; Accepted: 8 August 2016; Published: 11 August 2016

Abstract: The results of a study of ultra-rapid (flash) sintering of oxide ceramic materials under microwave heating with high absorbed power per unit volume of material ($10–500$ W/cm^3) are presented. Ceramic samples of various compositions—Al_2O_3; Y_2O_3; $MgAl_2O_4$; and $Yb(LaO)_2O_3$—were sintered using a 24 GHz gyrotron system to a density above $0.98–0.99$ of the theoretical value in $0.5–5$ min without isothermal hold. An analysis of the experimental data (microwave power; heating and cooling rates) along with microstructure characterization provided an insight into the mechanism of flash sintering. Flash sintering occurs when the processing conditions—including the temperature of the sample; the properties of thermal insulation; and the intensity of microwave radiation—facilitate the development of thermal runaway due to an Arrhenius-type dependency of the material's effective conductivity on temperature. The proper control over the thermal runaway effect is provided by fast regulation of the microwave power. The elevated concentration of defects and impurities in the boundary regions of the grains leads to localized preferential absorption of microwave radiation and results in grain boundary softening/pre-melting. The rapid densification of the granular medium with a reduced viscosity of the grain boundary phase occurs via rotation and sliding of the grains which accommodate their shape due to fast diffusion mass transport through the (quasi-)liquid phase. The same mechanism based on a thermal runaway under volumetric heating can be relevant for the effect of flash sintering of various oxide ceramics under a dc/ac voltage applied to the sample.

Keywords: microwave sintering; flash sintering; oxide ceramics; electric conductivity; grain boundary melting; densification

PACS: 81.40.Wx

---

1. Introduction

In recent years considerable interest has been drawn to the processes of materials sintering making use of electric currents and/or fields. Enhanced sintering of various ceramic, composite and metal powder materials has been observed when using such methods as Field Assisted Sintering Techniques (FAST), Pulsed Electric Current Sintering (PECS), Spark Plasma Sintering (SPS), and Microwave Sintering. Several reviews of these methods and their applications to the sintering of a wide range of different ceramics have been published recently (see, e.g., [1–3]). These new techniques have attracted great attention due to their common advantage, viz. a shorter time needed to consolidate powders as compared to conventional methods. In many cases, the reduction in the sintering time can be as large as 10^2, and the total processing time can be minutes instead of hours.

Recently, an even faster sintering method has been developed, the so called flash sintering [4–8]. In this method, a DC or low-frequency AC voltage is applied to a ceramic powder compact heated in a conventional furnace. The flash sintering occurs at a certain critical combination of the values of the temperature and the power dissipated in the sample due to the flow of electric current through it, and it results in full densification of the compact in a few seconds. The mechanisms responsible for flash sintering have not yet been determined, being a subject of wide discussion. Several hypothetic concepts supposedly relevant to flash sintering have been listed in the review [2]. They include nucleation of Frenkel pairs under the applied field [9], electrical boundary resistance [10], stress-induced generation of electric fields [11], and interaction between the external field and the space charge field [12]. Recently, a liquid film capillary model was discussed as a mechanism of rapid densification during flash sintering [13]. A similar mechanism of grain boundary softening has been proposed to explain the rapid densification under SPS [14]. In fact, starting from the very first publications on the flash sintering effect observations it has been noted that the current flow through the sample should lead to its significant heating [7]. The internal Joule heat sources provide volumetric heating, whereas the removal of heat proceeds through the surface. It is clear that at a certain ratio between the capacity of energy generation and heat capacity of the sample an overheating instability, known as the thermal runaway, may develop. Currently, it is generally agreed that the temperature instability plays a crucial role in the development of the flash sintering effect. At the same time, there is no clear understanding of how/whether the temperature instability can be a trigger for fast densification.

The thermal runaway is a well-known issue in microwave processing of materials [15,16]. Microwave heating occurs due to the absorption of electromagnetic radiation in the material. In many materials, including ceramics of various compositions, the coefficients of absorption increase with temperature. If during the heating process the parameter $\beta = (dP_v/dT) \times (T/P_v)$ (where P_v is the power deposited per unit volume of the sample and T is temperature) exceeds a certain critical value, a thermal runaway develops in the sample [16]. As a rule, the possibility of uncontrolled temperature instability development is viewed as one of the main shortcomings of the use of microwave heating for high-temperature processing of materials.

We recently demonstrated [17,18] that flash sintering of oxide ceramic materials (Al_2O_3, Y_2O_3, $MgAl_2O_4$, and $Yb:(LaY)_2O_3$) is observed under rapid microwave heating due to the development of a temperature instability. There is experimental evidence that the ultra-rapid densification (within tens of seconds) of powder compacts to a near-theoretical density occurs due to particle surface softening and subsequent liquid phase sintering. In the development of the flash sintering process a determining role is played by two factors associated with the properties of the boundary regions of the grains. Due to the elevated concentration of defects and impurities in the grain boundaries these regions have higher electric conductivity (which grows fast as the temperature increases) and therefore higher absorption coefficient of microwave radiation. The increase in the absorbed power causes further increase in the temperature and in the effective electric conductivity. Although the power is deposited in a non-uniform manner because of a non-uniform distribution of the heat sources, thermal conduction effectively equalizes the temperature on the scale of the grain size (on the order of one micron or less) [19]. On the other hand, at temperatures that are close enough to the melting point the intense electromagnetic field may generate additional defects in the crystalline lattice. Due to the elevated defect/impurity concentration the temperature of the particle surface softening/pre-melting can be noticeably different from the melting point of the crystalline material of the bulk of the grains. Therefore it can be argued that intense microwave heating leads to formation of a (quasi-)liquid phase at the grain boundaries, which results in very rapid liquid-phase sintering. A proper fast feedback-based control of the microwave power during sintering helps prevent possible negative effects associated with the thermal runaway.

In this paper we discuss in detail the experimental results that confirm formation of the liquid phase in the samples sintered under intense microwave heating. The estimated values of the electric field, effective electric conductivity, and specific absorbed power are shown to be close to the respective values obtained in the flash sintering processes carried out under an applied DC or low-frequency AC electric field. Based on this correlation, we conclude that the mechanisms responsible for the flash sintering effect are similar for the DC/AC electric field-assisted processes and microwave sintering.

2. Materials and Methods

2.1. Materials Preparation and Characterization

Nanopowders of Y_2O_3, $MgAl_2O_4$, and 5 at % $Yb:(La_{0.1}Y_{0.9})_2O_3$ compositions were produced at the Institute of Chemistry of High-Purity Substances (Russian Academy of Sciences, Nizhny Novgorod, Russia) via a wet chemical route with a subsequent self-propagating high-temperature synthesis (SHS). The details of the powder production procedures can be found elsewhere [17,18]. The particle sizes of powders after calcining determined by the Brunauer, Emmett and Teller method (BET) were 130 nm (Y_2O_3), 10 nm ($MgAl_2O_4$), and 140 nm ($Yb:(LaY)_2O_3$). The Al_2O_3 samples were prepared from high purity α-alumina powder AES-11C (Sumitomo Chemical Co. Ltd., Tokyo, Japan). The average particle size D_{50} was 0.45 µm. According to the producer, 0.05 wt % MgO was intentionally introduced into the powder for better sinterability. The powders were cold uniaxially pressed at 150–400 MPa into disks, 13 mm in diameter and 2.5 mm in thickness. The relative densities of the green bodies were approximately 0.42 (Y_2O_3), 0.38 ($MgAl_2O_4$), 0.52 ($Yb:(LaY)_2O_3$), and 0.64 (Al_2O_3) of the corresponding theoretical densities.

The samples were heated in the applicator of a gyrotron system with a microwave power of up to 6 kW at a frequency of 24 GHz, equipped with a computerized feedback power control circuit [20]. The samples were placed in the center of a cylindrical quartz crucible, 100 mm in diameter and 100 mm in height. For thermal insulation of the samples the crucible was filled with either coarse (particle size 3–5 µm) Y_2O_3 powder (in the case of $Yb:(LaY)_2O_3$ and Y_2O_3 samples) or coarse alumina powder ($MgAl_2O_4$ and Al_2O_3 samples). The temperature of the samples was measured by a B-type thermocouple whose tip touched the sample center from the bottom. In the case of $Yb:(LaY)_2O_3$ sintering, a thin (0.5 mm) Y_2O_3 ceramic disk was placed between the sample and the thermocouple to prevent chemical interaction. To remove the residues of the binder, samples were heated in air at a slow heating rate (10 °C/min) up to an intermediate temperature of 800 °C and held for one hour. No alterations in density and microstructure of samples were observed after this preliminary heat treatment. Upon completion of this initial stage of heating, the applicator was pumped out to a pressure of about 5×10^{-1} Pa and then the main stage of fast heating to the preset maximum temperature started. The Al_2O_3 powder compacts were microwave sintered in air at normal pressure.

The samples were heated from the above mentioned intermediate temperature to the maximum sintering temperature in two different regimes. One employed computer control of the microwave power enabling the heating of a sample at a preset fixed rate (50, 100, 150, or 200 °C/min). In the other regime, a fixed level of microwave power was used (about 5 kW as measured at the applicator input). In both cases the heating was terminated when the preset maximum temperature was reached. The microwave power was switched off automatically by the control system and the sample cooled down along with the thermal insulation surrounding it.

For the comparative study of the grain growth kinetics the samples were heated by microwaves and conventionally at a ramp-up rate of 6 °C/min to a number of temperatures in the range 1320–1770 °C and held at these temperatures for zero and ten hours. Mean linear intercepts, *L*, were measured on SEM images of thermally etched polished surfaces containing 50–150 grains, and converted to grain size, *D*, taking into account the stereological factor [21].

A key point in the comparative studies is the accordance between the results of temperature measurements at microwave and conventional heating. An unshielded B-type Pt–Rh thermocouple was used to measure the temperature of samples at microwave heating. The correctness of the thermocouple measurements was checked by microwave heating of a small copper ball making use of the Cu melting temperature (1083 °C) as a reference point. The results of tests have shown that the difference between the measured melting temperature and the reference data did not exceed 5 °C, which was within the accuracy of the thermocouple measurements.

The density of the sintered samples was determined by Archimedes weighing in distilled water with an estimated accuracy of ±0.01 g/cm^3. The microstructure of the as-sintered samples was studied by scanning electron microscopy (JEOL JSM-6390 LV, Tokyo, Japan). The element distribution was analyzed by energy dispersion spectrometry (EDS) using JEOL EX-54175 JMH (Tokyo, Japan) detector combined with the microscope. The grain structure profiles were investigated by atomic force microscopy (Smena NT–MDT, Moscow, Russia). The phase composition of the sintered samples was analyzed using a Rigaku Ultima IV X-ray difractometer (Tokyo, Japan).

2.2. Energy Balance during Microwave Heating

As noted above, the connection between the development of the overheating instability (thermal runaway) and flash sintering is now commonly recognized. The instability develops due to a misbalance between the power deposited within the volume of the sample and the heat losses. Therefore a key parameter determining the flash sintering phenomenon is the power deposited per unit volume. While in the experiments with DC/low-frequency AC currents the power density is easily obtainable by measuring the current flowing through the sample and the applied voltage, in the case of microwave heating the power absorbed per unit volume is not readily available but can be determined using the procedure described below.

The microwave power absorbed in a sample increases its temperature and/or compensates the heat losses. The microwave power deposited per unit volume of the sample, P_v, can be determined from the experimental data using the energy balance equations. Immediately before and after the time instant when the maximum temperature is achieved and the microwave power is switched off the energy balance equations take the following form:

$$P_v = \rho C \left(\frac{dT}{dt}\right)_+ + P_{hl}, P_{hl} = \rho C \left|\left(\frac{dT}{dt}\right)_-\right| \tag{1}$$

where ρ is density, C is specific heat capacity of the material of the sample, P_{hl} is the power in heat losses, $(dT/dt)_+$ and $|(dT/dt)_-|$ are the rates of heating and cooling before and after the microwave power switchoff, respectively. As follows from Equation (1), the value of P_v is higher, the higher the heating rate (as long as the thermal insulation conditions are the same). Therefore it can generally be argued that whenever the experimental results depend on the microwave heating rate, it is in fact the effect of the absorbed electromagnetic power.

In general, the power in heat losses, P_{hl}, is the total power lost by thermal conduction, convective and radiative heat flows. The power in heat losses increases as the temperature of the sample grows. Therefore, during a constant-rate heating process the microwave power deposited per unit volume of the sample increases monotonically to compensate the increasing heat losses (unless there are phase transformations accompanied with the changes in the internal energy of the sample). An example of such behavior of microwave power during the heating of a sample compacted from α-Al$_2$O$_3$ powder at a rate of 10 °C/min to 1100 °C and then 15 °C/min to 1400 °C is shown in Figure 1.

The power P_v (1) absorbed in the sample, per unit volume, can be expressed via the electric field strength in the sample, E_s, and the effective high-frequency electric conductivity of the material of the sample, σ_{eff}, as follows:

$$P_v = \sigma_{\text{eff}} E_s^2 \tag{2}$$

Unlike the experiments on flash sintering with dc/low-frequency ac currents, in the microwave sintering experiments it is not possible to directly measure the power absorbed in the volume of the sample. The measureable quantity in these experiments, along with the temperature of the sample, is the input microwave power, P, that is fed into the applicator in which the sample and the thermal insulation arrangement surrounding it are positioned. The input power, P, determines the magnitude of the electromagnetic energy, W, stored in the cavity [22]:

$$W = \varepsilon_0 \int E^2 dV = \frac{Q}{2\pi f} P \qquad (3)$$

where ε_0 is the electric constant, E is the electric field in the cavity, Q is the quality factor of the cavity, and f is the microwave frequency.

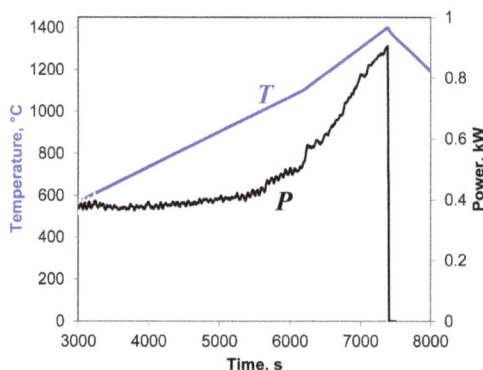

Figure 1. Temperature (T) and microwave power (P) during microwave heating of a sample compacted from α-Al$_2$O$_3$ powder at a rate of 10 °C/min to 1100 °C and then 15 °C/min to 1400 °C.

A specific feature of the gyrotron system that operates in the millimeter-wave range is that its applicator is an untuned multimode cavity [23] with a high ratio of its volume, V, to the cube of the radiation wavelength, λ: $V/\lambda^3 \approx 10^5$. In such a cavity the electromagnetic field distribution over its volume is quasi-uniform since it presents a superposition of many oscillation modes excited simultaneously [24]. The quality factor, Q, for this type of cavity is determined primarily by the Ohmic losses in its walls, because the losses in the sample are small due to its small dimensions (usually on the order of 1 cm^3) and the coupling loss is negligible due to small area of the input opening of the cavity.

It can be shown that the electric field strength E_s in a small sample with the dielectric permittivity $\varepsilon \sim 5$–15, which is placed in the cavity, is approximately equal to the electric field in the cavity outside of the sample: $E_s \approx E$. Then, using Equations (2) and (3) it is easy to obtain the interrelation between the power incoming the applicator, P, and the microwave power deposited per unit volume of the sample:

$$P_v \approx \frac{\sigma_{eff}}{2\pi\varepsilon_0 f} Q \frac{P}{V} \qquad (4)$$

where V is the volume of the applicator.

3. Results and Discussion

3.1. Flash Microwave Sintering of Yb:(LaY)$_2$O$_3$ Ceramic Samples

In the present study, at temperatures above 800 °C the Yb:(LaY)$_2$O$_3$ samples were heated at rates in the range of 50–7000 °C/min up to a preset maximum temperature chosen between 1300 and 1600 °C. The cooling rates (immediately after the automatic microwave power switchoff at maximum temperature) were 180–1000 °C/min, depending on the heating regime. An example of a typical behavior of the input microwave power, *P*, during these microwave heating experiments is shown in Figure 2.

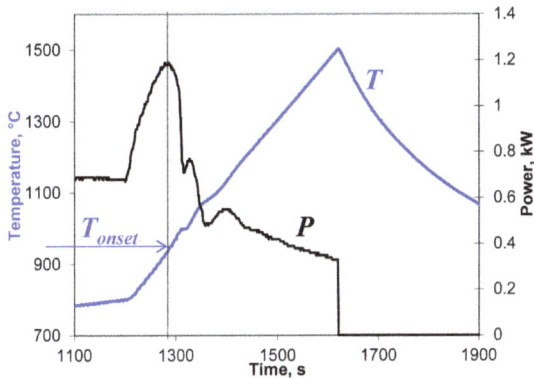

Figure 2. Temperature and microwave power at the applicator input vs. time, recorded at the high-temperature stage of the sintering of an Yb:(LaY)$_2$O$_3$ sample with a heating rate of 100 °C/min. The vertical line denotes the onset of the sharp rise in the effective conductivity.

As seen in Figure 2, at an initial stage of heating the automatically controlled input microwave power increases from 680 to 1200 W as the temperature grows. Then, at a temperature of about 950 °C, the actual temperature growth rate begins to exceed the prescribed one. The accelerated temperature growth, i.e., thermal runaway, is suppressed by the automatic process control system by sharply reducing the microwave power. As follows from Equation (4), the drop in the power needed to sustain the heating at a constant rate means nothing but a sharp increase in the effective electric conductivity of the sample. The experiments show that the temperature T_{onset} at which the power drop occurs decreases with increasing heating rate [17]. Obviously, the observed increase in the effective conductivity must be caused by certain changes in the structure or phase state of the material, which we will identify below. Let us note that the detection of such changes based on the measurement of the input microwave power can be viewed as a peculiar implementation of the method of microwave thermal analysis [25].

The final density of all flash microwave sintered Yb:(LaY)$_2$O$_3$ samples was 98%–99% [17]. An SEM study of unpolished surfaces of the samples showed that the heating rate (and hence the absorbed microwave power as discussed above) strongly affects the microstructure. Isolated droplets, a fraction of a micron in size, located along the grain boundaries are observed in the sample heated at a rate of 50 °C/min to a temperature of 1500 °C (Figure 3a). In previous studies with samples of the same composition, microwave heated at a much slower rate of 5 °C/min [26], no such droplets were seen in the microstructure. With an increase in the microwave heating rate to 100 and 150 °C/min, the particles merge together forming layers, up to 0.3 µm in thickness, which surround the grains (Figure 3b). Between the grains one can clearly see interlayers composed of a different phase (looking brighter in the backscattered electron image). The formation of rounded or lenticular islands at grain boundaries that first produce a necklace structure and then aggregate into continued films is a well-known effect in the theory of liquid-phase sintering [27].

The thickness of the melted intergranular layers varies with the heating rate and the maximum temperature of the sample. If both the heating rate and the maximum temperature were excessively high, the melting process was not confined within the grain boundaries but affected larger (though still localized) areas of the sample (Figure 3c). An XRD study of the surfaces of all sintered samples, including those with partially melted regions, revealed only crystalline and no amorphous phases.

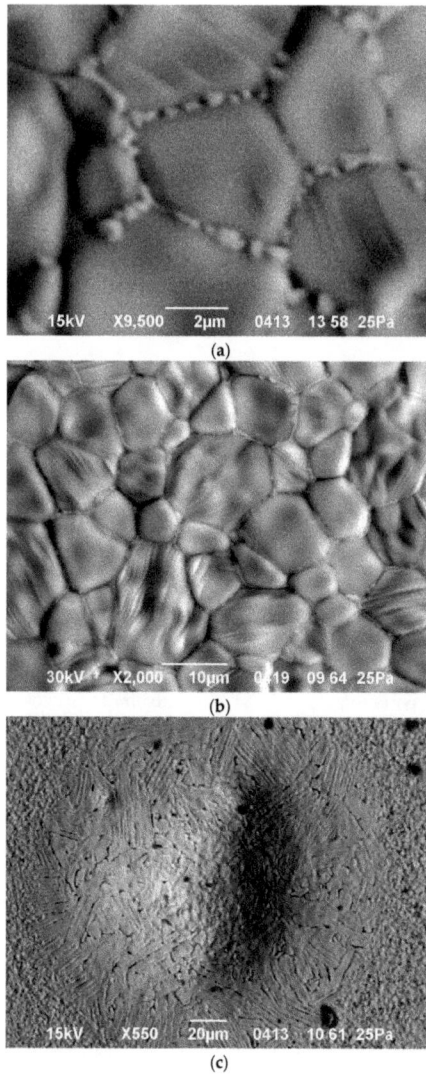

Figure 3. Microstructure of unpolished surfaces of sintered Yb:(LaY)$_2$O$_3$ samples heated in the regimes: (a) 50 °C/min to 1500 °C; (b) 100 °C/min to 1500 °C; (c) 7000 °C/min to 1580 °C.

The presence of a liquid phase between solid particles was confirmed by SEM and AFM images of the surfaces of the sintered ceramic samples. For example, an SEM study of the sample microwave heated at a rate of 200 °C/min to 1500 °C showed that the surface relief is uneven on the grain size scale (Figure 4a). The edges of the grains are raised above their middle area. The results of an AFM study of the surface relief confirmed that the grain edges are protruding (Figure 4b). Typically, the height of these edge ledges is of the same order of magnitude as the width of the softened intergranular layers, i.e., a fraction of a micron. The protruding areas arise at grain boundaries because the volume of material changes as the liquid phase forms. In most oxide materials the specific volume increases by 10%–20% upon transition from solid to liquid phase [28]. During sintering, the liquid phase partially fills the triple points of the granular structure by viscous flow, thus facilitating densification, and is partially extruded to the free surface of the samples.

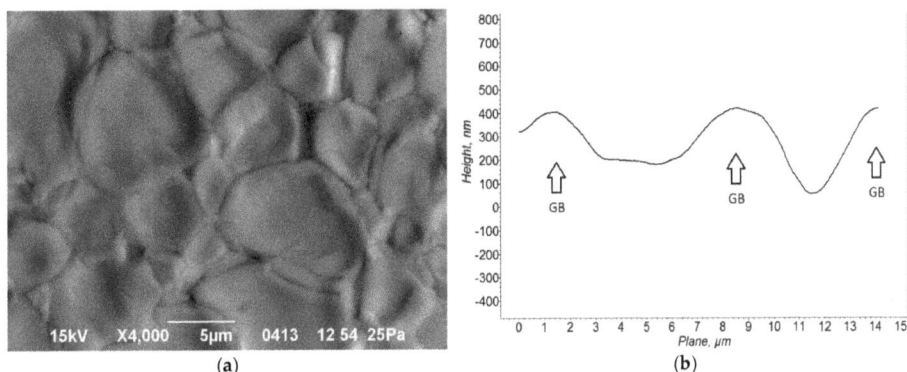

(a)

(b)

Figure 4. Microstructure of an unpolished surface of a sintered Yb:(LaY)$_2$O$_3$ sample microwave heated at a rate of 200 °C/min to 1500 °C: (**a**) SEM image; (**b**) AFM profile across two adjacent grains, grain boundary positions shown with arrows (note different scales for the horizontal and vertical axes).

The effect of accelerated sintering apparently results from an avalanche pre-melting or melting of powder particle surfaces [29,30]. The non-uniform deposition of microwave energy starts at an early stage of heating due to enhanced absorption of microwave energy at the particle surfaces, where the concentration of impurities and defects is elevated. Despite this non-uniformity in energy deposition, the temperature remains nearly uniform across the particle because thermal conductivity prevents development of a significant difference between the temperatures of the particle surface and bulk when the particle size is on the order of microns [19]. However, due to the abundance of impurities and defects in the near-boundary regions of particles the melting temperature of surface/boundary can differ noticeably from the melting point of the pure solid material. As a result, particle surface pre-melting may occur well below the bulk melting point of the material. This leads, in turn, to a sharp increase in the effective conductivity σ_{eff} and the absorbed microwave power P_v, causing the development of a local thermal runaway. Due to the melting of particle surfaces the solid grains appear to be surrounded by a melt with a low viscosity. Dissolution of solid into the liquid and enhanced diffusion mass transport through the liquid layer leads to a rounded shape and a smooth surface of grains.

The liquid phase wets the grains completely because their chemical compositions are similar. The capillary attractive force causes particle rearrangement due to rotation and sliding of small-size grains relative to each other, which eventually results in fast densification. At the final stage of sintering the larger grains grow at the expense of the smaller grains by the solution-reprecipitation mechanism. As seen in the microstructure of an unpolished surface of a sample heated up to a temperature of 1570 °C (Figure 5), some of the larger-size (about 20 µm) grains have an inner substructure that consists of densely packed rounded particles of an order-of-magnitude smaller size. Note that a similar microstructure, with grains having an inner substructure, was observed in the YAG ceramics sintered by spark plasma sintering [31], which was interpreted as a manifestation of rapid densification of nanocrystalline YAG via surface softening of particles and liquid phase formation.

Figure 5. Microstructure of an unpolished surface of a sintered Yb:(LaY)$_2$O$_3$ sample microwave heated at a rate of 100 °C/min to 1570 °C. Note the substructure in some of the larger grains.

It is well known that during microwave volumetric heating, accompanied by heat loss through the surface, the so-called inverse temperature distribution develops in the sample, with the core of the sample being hotter than the periphery [16]. For example, during microwave heating at a rate of 150 °C/min to 1500 °C the temperature difference between the center and the surface of an Yb:(LaY)$_2$O$_3$ sample with a diameter of 13 mm reached 250–300 °C [17]. Yet, despite such a large temperature difference, the microwave flash sintered samples achieved uniform near-full density, and the grain size distribution over the diameter of the samples was fairly uniform, with deviations from the average value not exceeding ±10%. Based on the analysis of the experimental observations described above, the following mechanism of microwave flash sintering can be suggested. The process of particle surface melting starts in the core region of the sample. In the course of densification the liquid phase is partially squeezed out of the core region into the more porous peripheral structure. The region of the maximum deposition of the microwave power, P_v, moves toward the periphery along with the liquid phase due to the elevated electric conductivity of the latter. This results in further melting of grain boundaries and liquid phase production outside of the core region of the sample. In this manner, a transient liquid-phase densification front propagates from the core of the sample to its periphery, resulting in high final density and uniform grain size distribution. It should also be noted that the liquid-phase front propagation limits the development of local thermal runaway at each particular point of the sample and thereby prevents its destruction.

An EDX study has shown that the initially uniform element distribution corresponding to the stoichiometric compound 5 at % Yb:(La$_{0.1}$Y$_{0.9}$)$_2$O$_3$ changes after rapid sintering. The characteristic element composition of the phases was determined by averaging the results of measurements at five points belonging to the intragrain areas and seven points in the intergranular phase. Each measurement characterized a spot on the order of 1 μm in size. The measurements were performed on unpolished surfaces of the samples to increase the sensitivity to the intergranular phase that squeezed out from between the grains and distributed partially over the surface of the sample. The results of element analysis, including the values of the experimental error, are listed in Table 1. A histogram of the element content in the intragrain and intergrain regions is shown in Figure 6.

Table 1. Results of element analysis, wt %.

Element	Intragrain	Intergranular Phase
C	1.80 ± 0.09	1.50 ± 0.06
O	14.44 ± 0.47	17.36 ± 0.81
Y	68.83 ± 0.37	59.57 ± 2.00
La	9.13 ± 0.14	16.53 ± 1.70
Yb	5.79 ± 0.10	5.04 ± 0.18

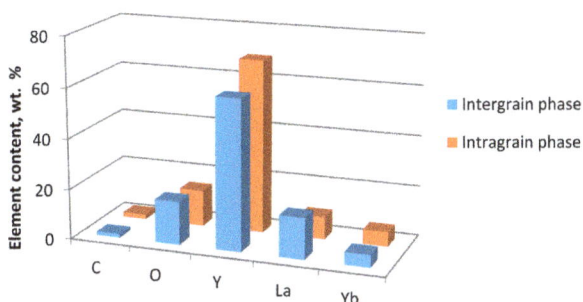

Figure 6. Results of element analysis on the surface of a sintered Yb:(LaY)$_2$O$_3$ sample microwave heated at a rate of 100 °C/min to 1550 °C.

Compared to the material of the grains, the intergranular phase is enriched with La and O and depleted with Y. This has also been confirmed qualitatively by element maps obtained by X-ray microanalysis. The relative element content in the bulk of the grains is close to the stoichiometric ratio for the composition 5 at % Yb^{3+} (La$_{0.1}$Y$_{0.9}$)$_2$O$_3$; however, in the intergranular phase it is notably different. Presumably, the La^{3+} ions have a lower diffusion coefficient than Y^{3+} due to their larger mass and ionic radius; therefore La ions may accumulate in the intergranular regions.

3.2. Microwave Effect on the Grain Growth Kinetics of Yb:(LaY)$_2$O$_3$ Ceramics

It is known that the Y$_2$O$_3$—(0–18 mol %) La$_2$O$_3$ solid solution with a cubic structure is thermodynamically stable at temperatures above 1400 °C [32–34]. No thermodynamic data on the 5 at % Yb^{3+} (La$_{0.1}$Y$_{0.9}$)$_2$O$_3$ composition, in particular on its melting temperature, are available. In order to study the effect of the microwave electromagnetic field on grain boundary melting phenomena, liquid phase formation, and its influence on grain growth, samples of this composition were heated both by microwaves and conventionally (in a resistive oven). This experimental series was performed at a ramp-up rate of 6 °C/min to make possible the comparison between the results of microwave and conventional heating.

An example of a typical behavior of the input microwave power during the heating at a rate of 6 °C/min is shown in Figure 7. The vertical line denotes the onset of the rise in the effective conductivity presumably caused by the melting of particle surfaces. At this low heating rate, the onset temperature, T_{onset}, is 1140 ± 20 °C, which coincides within the margin of error with the value 1130 ± 20 °C observed at a heating rate of 50 °C/min in the experimental series described in Section 3.1. This suggests that the temperature of particle surface softening in the presence of moderate-intensity microwave field is about 1140 °C, decreasing at higher field intensities.

It is well known that the process of grain growth is greatly accelerated in the presence of the liquid phase in the sintered material [35,36]. The rate of diffusion mass transport is proportional to the product of the diffusion coefficient and the cross section of the diffusion layer. At liquid phase sintering the magnitude of this product is high compared with the case of solid phase sintering due to an increase in both factors. As a result, the rates of both densification and grain growth are higher at liquid phase sintering. The features related to microwave radiation absorption and microwave interaction with the material may develop at any of the main consecutive stages of liquid-phase sintering [36]: (a) melting of the liquid-forming additive and redistribution of the liquid; (b) rearrangement of the majority solid phases in the presence of a liquid phase; (c) densification and shape accommodation of the solid phase; and (d) final densification driven by residual porosity in the liquid phase.

Therefore, a comparative study of the grain growth kinetics should make it possible to understand the mass transport mechanisms involved in the sintering process. The grain growth was systematically studied under microwave and conventional resistive heating by varying the temperature and hold time.

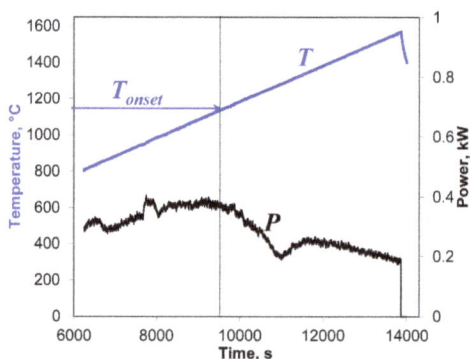

Figure 7. Temperature and microwave power at the applicator input vs. time, recorded at the high-temperature stage of the sintering of an Yb:(LaY)$_2$O$_3$ sample with a heating rate of 6 °C/min. The vertical line denotes the onset of the rise in the effective conductivity.

The grain growth by different mechanisms of diffusion mass transport is described by the following expression [35]:

$$D^m(t) - D^m(t_0) = Kt \tag{5}$$

where $D(t)$ and $D(t_0)$ are the grain sizes at time t and t_0, respectively, m is the grain size exponent,

$$K = K_0 \exp(-Q_a/RT) \tag{6}$$

K_0 is the pre-exponential factor of the diffusion coefficient, Q_a is the grain growth activation energy, and R is the gas constant. The exponent m depends on the rate-controlling mechanism of grain growth. The value $m = 2$ indicates that the solid state mass transport is the dominant mechanism for grain growth, whereas $m = 3$ is indicative of the diffusion though the liquid phase [37] or ion dissolution as the rate-controlling step.

The average grain size was determined using the SEM images obtained on the polished surfaces of the samples. The microstructure of samples sintered under microwave and resistive heating at a temperature of 1750 °C with a 10-h isothermal hold is shown in Figure 8. Figure 9 shows the dependencies of the grain size (averaged across the diameter of the sample) on the sintering temperature at microwave and conventional heating for zero and 10-h hold times.

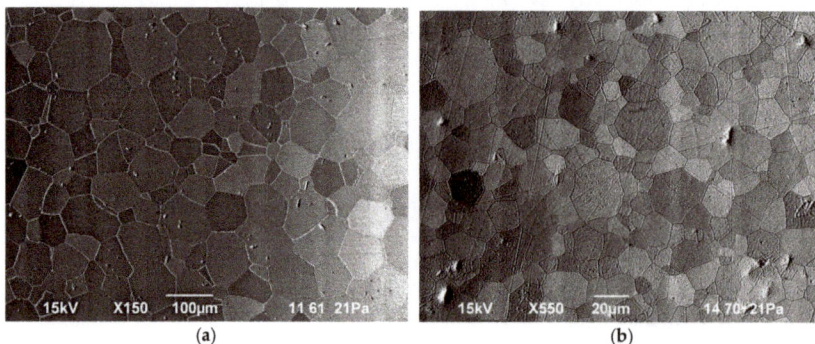

Figure 8. Microstructure of polished surfaces of sintered Yb:(LaY)$_2$O$_3$ samples heated at a rate of 6 °C/min to 1750 °C with a 10-h hold under (**a**) microwave, and (**b**) conventional heating. Note different scale bars.

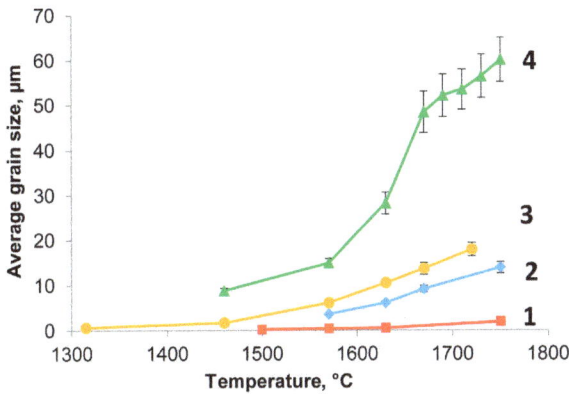

Figure 9. Average grain size in Yb:(LaY)$_2$O$_3$ samples vs. sintering temperature: 1—conventional heating, zero hold; 2—conventional heating, 10-h hold; 3—microwave heating, zero hold; 4—microwave heating, 10-h hold.

As seen from the data plotted in Figure 9, the heating method affects the grain growth rate substantially. At the same temperature and hold time, the average grain size obtained under microwave heating exceeds the grain size obtained under conventional heating greatly. For example, at a temperature of 1570 °C and zero hold the average grain size is 0.52 μm under conventional heating and 6.1 μm under microwave heating; at a temperature of 1750 °C and a 10-h hold the average grain size is 13.9 μm and 60.1 μm, respectively.

In the case of microwave heating, the character of the dependency of the average grain size on the temperature and hold time is typical of liquid phase sintering [38]. The rapid growth of the average grain size at intermediate temperatures (1550–1650 °C) corresponds to the predominant effect of the solution-reprecipitation mechanism at this stage. At higher temperatures the grain growth slows down and the microstructure coarsening by the Ostwald ripening mechanism can become the dominant process. This grain growth slowdown is as well characteristic of conventional and spark plasma liquid phase sintering [39,40] and it does not depend on the method of heating.

The activation energy of grain growth can be determined on the basis of the obtained data using Equation (5). Plotted in Figure 10 are the dependencies of the quantity $\ln[(D^m(t) - D^m(t_0)]$ on the reciprocal temperature. The values of the activation energies and the coefficients of determination corresponding to a straight line fit of the Arrhenius dependence of $\ln[(D^m(t) - D^m(t_0)]$ on T^{-1} are listed in Table 2 for different m values.

Table 2. Activation energy of grain growth and the coefficient of determination, R^2, for the dependence of $\ln[(D^m(t) - D^m(t_0)]$ on the reciprocal temperature of sintering for microwave and conventional heating.

Method of Heating	Temperature (°C)	Activation Energy, Q_a (kJ/mol)/ Coefficient of Determination, R^2	
		$m = 2$	$m = 3$
microwave	1570–1670	723/0.992	1056/0.994
	1670–1750	147/0.9725	241/0.980
conventional	1570–1750	467/0.983	706/0.984

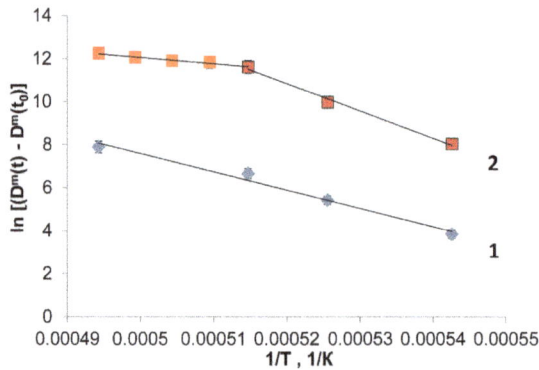

Figure 10. Plots of $\ln[(D^m(t) - D^m(t_0)]$ vs. reciprocal temperature of sintering for conventional (**1**) and microwave (**2**) heating; here it is assumed that $m = 3$.

One or another mechanism controlling the grain growth could be chosen based on the comparison of the obtained values of the activation energy for grain growth with the known data on the activation energy for diffusion processes in the given material in the appropriate temperature range. Unfortunately, there are no data on the activation energy for diffusion available for the Y_2O_3—10 mol % La_2O_3 (doped with 5 at % Yb) solid solution.

It can be seen from the data listed in Table 2 that for the case of conventional heating the R^2 values for both fits ($m = 2$ and $m = 3$) virtually do not differ. The hypothesis of $m = 2$ appears preferable because in this case the corresponding value of Q_a is closer to the typical values of the activation energy of grain boundary diffusion of Y^{3+} ions in solids [41,42]. In the case of microwave heating, the plot of $\ln[(D^m(t) - D^m(t_0)]$ shown in Figure 10 has two distinct parts for low (1570–1670 °C) and high temperatures (1670–1750 °C). This behavior is typical of liquid phase sintering due to the different mechanisms of grain growth acting at low and high temperature. The calculated values of the activation energy differ greatly for the two parts of the plot: for $m = 3$ the activation energy $Q_a = 241$ kJ/mol for the temperature range 1570–1670 °C and $Q_a = 1056$ kJ/mol for the range 1670–1750 °C. Large activation energy values for the high-temperature range are in good agreement with the activation energy for dissolution of rare-earth oxides which is above 1000 kJ/mol [43].

3.3. Ultra-Rapid Microwave Sintering of Al_2O_3, $MgAl_2O_4$, and Y_2O_3 Ceramics

The observed flash microwave sintering effect is not unique to Yb:$(LaY)_2O_3$ laser ceramics. The heating procedure, conditions and characteristic features of ultra-rapid microwave sintering have been studied for other oxide ceramics: Y_2O_3, Al_2O_3, and $MgAl_2O_4$ [18,44]. Similar to the case of Yb:$(LaY)_2O_3$ ceramics, the occurrence of the microwave flash sintering effect was in all cases determined by the combined action of two factors: the temperature of the sample and the microwave power deposited per unit volume of the sample, P_v. The overheating of the samples was avoided and their integrity was ensured by a system of fast automatic process control, which prevented development of the thermal runaway. From the practical viewpoint, it is important that the onset of rapid densification is easily identified without dilatometry by a sharp decrease in the level of the input microwave power required to sustain the preset heating rate.

The Y_2O_3 samples were microwave heated at rates 10–200 °C/min to maximum temperatures 1400–1700 °C with zero hold time. Both a sharp drop in the input microwave power due to an increase in the material's effective conductivity and a simultaneous increase in the density of the samples were observed at temperatures 1350–1450 °C when the microwave power deposited per unit volume of the sample, P_v, was above 40 W/cm³. As an example, shown in Figure 11 are the temperature and the microwave power at the applicator input recorded at the high-temperature stage of the sintering of an Y_2O_3 sample with a heating rate of 100 °C/min to a maximum temperature of 1600 °C. Y_2O_3 samples with densities over 0.98 ρ_{th} have been obtained at a heating rate of 100 °C/min and maximum temperatures 1600–1700 °C. The density of the microwave flash sintered Y_2O_3 samples was higher than the density of the samples obtained by a slower-rate (10 °C/min) microwave heating process with a 15 min isothermal hold at the maximum temperature.

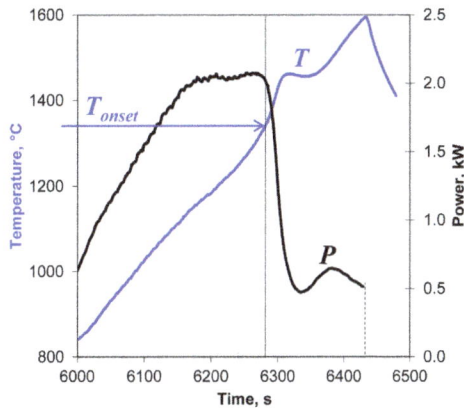

Figure 11. Temperature and microwave power at the applicator input vs. time, recorded at the high-temperature stage of the sintering of an Y_2O_3 sample with a heating rate of 100 °C/min. The vertical line denotes the onset of the sharp rise in the effective conductivity.

The Al_2O_3 powder compacts were microwave sintered in air at heating rates in the range 50–250 °C/min. The microwave power deposited per unit volume of the sample, P_v, was 15–100 W/cm³. Samples, with the density as high as 98%–99% of the theoretical density (ρ_{th}), were obtained at temperatures 1400–1550 °C with zero hold time. Additionally, Al_2O_3 samples were sintered in a preheated resistive furnace in the regimes which mimicked the regimes of microwave heating precisely, to make possible the comparison between the results of fast microwave and conventional sintering. The density of the samples sintered in the resistive furnace did not exceed 0.96 ρ_{th}.

The density of the $MgAl_2O_4$ samples microwave heated at rates in the range 100–200 °C/min to a maximum temperature of 1800 °C with zero hold time was over 0.99 ρ_{th}. This was markedly larger than the density of samples heated at a lower rate of 6 °C/min to the same temperature with a 2-h hold time (about 0.97 ρ_{th}).

It should be noted that in contrast to Yb:(LaY)$_2$O$_3$-samples, the samples of all these compositions were compacted from high purity powders and contained no intentionally introduced additives, except for 0.05% MgO in Al_2O_3. Due to this fact, only minor amounts of liquid phase could be formed which were not clearly detectable by the analytical instruments used in this study. However, the specific behavior of the microwave power during the heating of the above listed ceramic materials, as well as the common features in the grain growth and microstructure formation suggest that their ultra-rapid densification at high values of the microwave power deposited per unit volume, P_v, resulted from the same mechanism of fast mass transport via the softened particle surfaces as in the case of the Yb:(LaY)$_2$O$_3$ ceramic sintering.

3.4. Similarity between Microwave and DC/Low-Frequency AC Flash Sintering

The effect of flash sintering under an applied DC/AC voltage is usually discussed in terms of the electric field strength and the power absorbed per unit volume of the sample [4,7,29]. As was shown in [17], the observed microwave flash sintering effect can be discussed in similar terms.

The energy balance Equations (1) make it possible to estimate the microwave power absorbed per unit volume of the sample using the data on the rates of heating $(dT/dt)_+$ and cooling $(dT/dt)_-$ recorded in the experiments immediately before and after the microwave power switchoff. On variation of the microwave power input to the applicator from 0.5 to 3.0 kW, which corresponds to the variation of the heating rate from 50 to 2100 °C/min, the value of the power deposited per unit volume of the samples, P_v, ranges from approximately 20 to 160 W/cm^3. Note that the electric power density triggering the dc/ac flash sintering of various oxide ceramics is typically in the range 10–1000 W/cm^3 [4,7,29]. From the relationship Equation (3) between the electromagnetic energy, W, stored in the cavity and the input power, P, taking into account the quality factor for the applicator used in the experiments, $Q \approx 10^4$, the microwave electric field strength can roughly be estimated as $E[\text{V/cm}] \approx 5 \times \sqrt{P[\text{W}]}$. At the mentioned variation of the input power the microwave electric field strength in the samples varies from 100 to 270 V/cm. The electric field of the same order of magnitude is typically imposed on the samples in the experiments on DC/AC flash sintering.

Using Equation (2) it is possible to estimate the effective microwave electric conductivity, σ_{eff}, based on the experimental data. It has been shown in [17] that the electric conductivity of the intergranular liquid phase can be estimated based on the percolation theory [45] assuming that it is much higher than the bulk conductivity of the grains. In particular, for the case when a sample was heated at a ramp-up rate of 2400 °C/min to 1500 °C it was estimated that the mass of the melted material could be as high as 20% of the total mass of the sample. The electric conductivity of the liquid phase was estimated to be 0.25–0.7 $(\Omega \cdot \text{cm})^{-1}$. The high-temperature electric conductivity of the 5 at % Yb^{3+} doped $(\text{La}_{0.1}\text{Y}_{0.9})_2\text{O}_3$ composition is not known. However, it should be noted that the estimated values are slightly lower than the conductivity of pure Y$_2$O$_3$ above the melting point, which equals 0.9 $(\Omega \cdot \text{cm})^{-1}$ [46].

Thus, the values of the electric field strength and the power deposited per unit volume of the sample at which the flash sintering effect is observed are within the same ranges for the cases of microwave heating and the heating by DC/AC currents. The results of microstructure characterization of the sintered materials suggest that the mechanisms responsible for the flash sintering effect in the DC/AC electric field-assisted processes and microwave sintering are identical. The formation of isolated small, mostly rounded droplets located along the grain boundaries, similar to those described in Section 3.1, was observed in a study of the microstructure of BaCe$_{0.8}$Gd$_{0.2}$O$_{3-\delta}$ powder compacts heated conventionally under an applied ac electric voltage [6]. According to reference [6], the ac current flowing through the grain boundaries promotes grain welding via local Joule heating.

The fact that the flash sintering effect is perhaps more pronounced in the case of DC/AC field-assisted sintering could be associated with the significant difference in the geometrical configuration of the samples used in the experiments. This difference results in a different degree of temperature non-uniformity over the sample cross section. The samples used in the DC/AC flash sintering experiments were either dog bones with a cross section about 2 × 3 mm^2 or small pellets, 5–7 mm in diameter, whereas the diameter of the pellets used in the microwave flash sintering experiments described here was 13 mm.

4. Conclusions and Outlook

We report the results of a study of an ultra-rapid ("flash") microwave sintering process with oxide ceramic materials Al$_2$O$_3$, Y$_2$O$_3$, MgAl$_2$O$_4$, and Yb:(LaY)$_2$O$_3$ to densities 98%–99% of the theoretical value within minutes or even fractions of a minute, without the high-temperature hold stage. On the grounds of the analysis of experimental data (microwave power, heating and cooling rates) and microstructure characterization, we propose a mechanism of flash microwave sintering based on particle surface softening/melting.

At a certain point during rapid microwave heating of samples at a fixed rate (50, 100, 150, or 200 °C/min) a sharp drop of the input microwave power is observed. This drop in the input power is produced by the automatic process control system as a response to the rapid increase in the high-frequency electric conductivity, which is associated with the development of an overheating instability known as thermal runaway. The development of the instability depends on two factors: the temperature of the material and the microwave power deposited per unit volume. The onset temperature of the instability decreases with an increase in the microwave power deposited per unit volume of sample.

The results of the microstructure characterization of the sintered samples demonstrate that the ultra-rapid densification occurs due to particle surface softening and subsequent liquid phase sintering. The temperature of the softening/melting of particle surfaces/grain boundaries can be noticeably different from the bulk material melting point due to the elevated density of defects and impurities. During microwave volumetric heating the highest temperature arises in the core of the sample and the process of particle surface melting starts there. In the course of densification the liquid phase is extruded into a more porous peripheral structure and contributes to its densification. In effect, a densification front, coinciding with the region of the maximum deposition of the microwave power, propagates from the core of the sample to its periphery, producing a fully dense ceramic material in a very short time.

A comparison of the obtained results of microwave flash sintering with the results of DC or low-frequency AC flash sintering experiments suggests that the mechanisms of ultra-rapid densification are similar or even identical for the two approaches. The temperature distribution in the samples subjected to a DC/AC electric field is determined by volumetric Joule heating and the surface heat loss, being thereby similar to the microwave heating case. The estimates of the specific deposited power, electric field strength, and electric conductivity give similar values of these quantities for the cases of DC/AC and microwave flash sintering.

From the applications standpoint, it should be emphasized that an undeniable advantage of the microwave flash sintering process is the fact that no electrodes are needed to supply the power to the articles undergoing sintering. It is also important that in the case of microwave sintering the use of a relatively simple fast process control system can be instrumental in preventing possible damage resulting from thermal runaway.

Microwave flash sintering is a novel process that offers significant advantages over conventional sintering methods in terms of process duration and energy consumption. Its further development should be aimed at the study of such factors as the properties of powder materials, optimization of material composition, geometrical limitations (if any) imposed on the configuration of the sintered products, structural, functional properties, and performance of the materials obtained by microwave flash sintering.

Acknowledgments: This research was supported in part by Russian Foundation for Basic Research, grant # 16-08-00736.

Author Contributions: Y.V.B. conceived and designed the experiments, S.V.E., A.G.E. and I.V.P. performed the microwave sintering experiments, A.A.S. performed microstructure characterization and analytical studies, V.V.K. developed the fast microwave process control system, Y.V.B. and K.I.R. analyzed the data and wrote the paper.

Conflicts of Interest: The authors declare no conflict of interest. The funding sponsors had no role in the design of the study; in the collection, analyses, or interpretation of data; in the writing of the manuscript, and in the decision to publish the results.

References

1. Munir, Z.A.; Quach, D.V.; Ohyanagi, M. Electric current activation of sintering: A Review of the pulsed electric current sintering process. *J. Am. Ceram. Soc.* **2011**, *94*, 1–19. [CrossRef]
2. Raj, R.; Cologna, M.; Francis, J.S.C. Influence of externally imposed and internally generated electrical fields on grain growth, diffusional creep, sintering and related phenomena in ceramics. *J. Am. Ceram. Soc.* **2011**, *94*, 1941–1965. [CrossRef]

3. Guillon, O.; Gonzales-Julian, J.; Dargatz, B.; Kessel, T.; Schierning, G.; Rathel, J.; Herrmann, M. Field-assisted sintering technology/Spark plasma sintering; Mechanisms, materials, and technology developments. *Adv. Eng. Mater.* **2014**, *16*, 830–849. [CrossRef]
4. Cologna, M.; Prette, A.L.G.; Raj, R. Flash-sintering of cubic yttria-stabilized zirconia at 750 °C for possible use in SOFC manufacturing. *J. Am. Ceram. Soc.* **2011**, *94*, 316–319. [CrossRef]
5. Yoshida, H.; Sakka, Y.; Yamamoto, T.; Lebrun, J.-M.; Raj, R. Densification behavior and microstructural development in undoped yttria prepared by flash-sintering. *J. Eur. Ceram. Soc.* **2014**, *34*, 991–1000. [CrossRef]
6. Muccillo, R.; Muccillo, E.N.S.; Kleitz, M. Densification and enhancement of the grain boundary conductivity of gadolinium-doped barium cerate by ultra fast flash grain welding. *J. Eur. Ceram. Soc.* **2012**, *32*, 2311–2316. [CrossRef]
7. Raj, R. Joule heating during flash-sintering. *J. Eur. Ceram. Soc.* **2012**, *32*, 2293–2301. [CrossRef]
8. Naik, K.S.; Sglavo, V.M.; Raj, R. Field assisted sintering of ceramics constituted by alumina and yttria stabilized zirconia. *J. Eur. Ceram. Soc.* **2014**, *34*, 2435–2442. [CrossRef]
9. Cologna, M.; Francis, J.S.C.; Raj, R. Field assisted and flash sintering of alumina and its relationship to conductivity and MgO-doping. *J. Eur. Ceram. Soc.* **2011**, *31*, 2827–2837. [CrossRef]
10. Ghosh, S.; Chokshi, A.H.; Lee, P.; Raj, R. A huge effect of weak DC electrical fields on grain growth in zirconia. *J. Am. Ceram. Soc.* **2009**, *92*, 1856–1859. [CrossRef]
11. Stuerga, D. Microwave-materials interactions and dielectric properties, key ingredients for mastery of chemical microwave processes. In *Microwaves in Organic Synthesis*, 2nd ed.; Loupy, E., Ed.; Wiley-VCH Berlag GmbH & Co.: Weinheim, Germany, 2006; pp. 1–61.
12. Jamnik, J.; Raj, R. Space charge controlled diffusional creep: volume diffusion case. *J. Am. Ceram. Soc.* **1996**, *79*, 193–198. [CrossRef]
13. Chaim, R. Liquid film capillary mechanism for densification of ceramic powders during flash sintering. *Materials* **2016**, *9*. [CrossRef]
14. Marder, R.; Estournès, C.; Chevallier, G.; Chaim, R. Spark and plasma in spark plasma sintering of rigid ceramic nanoparticles: A model system of YAG. *J. Eur. Ceram. Soc.* **2015**, *35*, 211–218. [CrossRef]
15. Roussy, G.; Bennani, A.; Thiebaut, J.M. Temperature runaway of microwave irradiated materials. *J. Appl. Phys.* **1987**, *62*, 1167–1170. [CrossRef]
16. Bykov, Y.V.; Rybakov, K.I.; Semenov, V.E. High-temperature microwave processing of materials. *J. Phys. D Appl. Phys.* **2001**, *34*, R55–R75. [CrossRef]
17. Bykov, Y.V.; Egorov, S.V.; Eremeev, A.G.; Kholoptsev, V.V.; Rybakov, K.I.; Sorokin, A.A. Flash microwave sintering of transparent Yb:(LaY)$_2$O$_3$ ceramics. *J. Am. Ceram. Soc.* **2015**, *96*, 3518–3524. [CrossRef]
18. Rybakov, K.I.; Bykov, Y.V.; Eremeev, A.G.; Egorov, S.V.; Kholoptsev, V.V.; Sorokin, A.A.; Semenov, V.E. Microwave ultra-rapid sintering of oxide ceramics. In *Processing and Properties of Advanced Ceramics and Composites VII (Ceramic Transactions, vol. 252)*; Mahmoud, M.M., Bhalla, A.S., Bansal, N.P., Singh, J.P., Castro, R., Manjooran, N.J., Pickrell, G., Johnson, S., Brennecka, G., Singh, G., et al., Eds.; Wiley: Hoboken, NJ, USA, 2015; pp. 57–66.
19. Johnson, D.L. Microwave heating of grain boundaries in ceramics. *J. Am. Ceram. Soc.* **1991**, *74*, 849–850. [CrossRef]
20. Bykov, Y.; Eremeev, A.; Glyavin, M.; Kholoptsev, V.; Luchinin, A.; Plotnikov, I.; Denisov, G.; Bogdashev, A.; Kalynova, G.; Semenov, V.; et al. 24–84-GHz gyrotron systems for technological microwave applications. *IEEE Trans. Plasma Sci.* **2004**, *32*, 67–72. [CrossRef]
21. Horovistiz, A.L.; Frade, J.R.; Hein, L.R.O. Comparison of fracture surface and plane section analysis for ceramic grain size characterisation. *J. Eur. Ceram. Soc.* **2004**, *24*, 619–626. [CrossRef]
22. Jackson, J.D. *Classical Electrodynamics*; Wiley: New York, NY, USA, 1962.
23. Kremer, F.; Izatt, J.R. Millimeter-wave absorption measurements in low-loss dielectric using an untuned cavity resonator. *Int. J. Infrared Millim. Waves* **1981**, *2*, 675–694. [CrossRef]
24. Kimrey, H.D.; Janney, M.A. Design principles for high-frequency microwave cavities. In *Microwave Processing of Materials (Mater. Res. Soc. Symp. Proc. Vol. 124)*; Sutton, W.H., Brooks, M.H., Chabinsky, I.J., Eds.; Materials Research Society: Pittsburgh, PA, USA, 1988; pp. 367–372.
25. Parkes, G.M.B.; Barnes, P.A.; Charsley, E.L.; Bond, G. Microwave thermal analysis—A new approach to the study of the thermal and dielectric properties of materials. *J. Therm. Anal. Calorim.* **1999**, *56*, 723–731. [CrossRef]

26. Balabanov, S.S.; Bykov, Y.V.; Egorov, S.V.; Eremeev, A.G.; Gavrishchuk, E.M.; Khazanov, E.A.; Mukhin, I.B.; Palashov, O.V.; Permin, D.A.; Zelenogorskii, V.V. Yb:(YLa)$_2$O$_3$ laser ceramics produced by microwave sintering. *Quantum Electron.* **2013**, *43*, 396–400. [CrossRef]
27. German, R.M.; Suri, P.; Park, S.J. Review: Liquid phase sintering. *J. Mater. Sci.* **2009**, *44*, 1–39. [CrossRef]
28. Samsonov, G.V. *The Oxide Handbook*; Plenum: New York, NY, USA, 1973.
29. Downs, J.A.; Sglavo, V.M. Electric Field Assisted Sintering of Cubic Zirconia at 390 °C. *J. Am. Ceram. Soc.* **2013**, *96*, 1342–1344. [CrossRef]
30. Narayan, J. A new mechanism for field-assisted processing and flash sintering of materials. *Scr. Mater.* **2013**, *69*, 107–111. [CrossRef]
31. Chaim, R.; Marder-Jaeckel, R.; Shen, J.Z. Transparent YAG ceramics by surface softening of nanoparticles in spark plasma sintering. *Mater. Sci. Eng. A* **2006**, *429*, 74–78. [CrossRef]
32. Coutures, J.; Foex, M. Etude à haute tempèrature du diagramme d'equilibre du système formè par le sesquioxyde de lanthane avec le sesquioxyde d'yttrium. *J. Solid State Chem.* **1974**, *11*, 294–300. [CrossRef]
33. Yoshimura, M.; Rong, X.-Z. Various solid solutions in the systems Y$_2$O$_3$-R$_2$O$_3$ (K = La, Nd, and Sm) at high temperature. *J. Mater. Sci. Lett.* **1997**, *16*, 1961–1963. [CrossRef]
34. Chen, M.; Hallstedt, B.; Gauckler, L.J. CALPHAD modeling of the La$_2$O$_2$-Y$_2$O$_3$ system. *CALPHAD* **2005**, *29*, 103–111. [CrossRef]
35. Rahaman, M.N. *Ceramics Processing and Sintering*; Marcel Dekker: New York, NY, USA, 1995.
36. De Jonghe, L.C.; Rahaman, M.N. Sintering of Ceramics. In *Handbook of Advanced Ceramics*; Somiya, S., Aldinger, F., Spriggs, R.M., Uchino, K., Koumoto, K., Kaneno, M., Eds.; Elsevier: London, UK; San Diego, CA, USA, 2003; pp. 187–264.
37. Brook, R.J. Controlled grain growth. In *Ceramic Fabrication Processes (Treatise on Materials Science and Technology, Vol. 9)*; Wang, F.F.Y., Ed.; Academic Press: New York, NY, USA, 1976; pp. 331–364.
38. German, R.M. *Sintering Theory and Practice*; Wiley: New York, NY, USA, 1996.
39. Kochawattana, S.; Stevenson, A.; Lee, S.-H.; Ramirez, M.; Gopalan, V.; Dumm, J.; Catillo, V.K.; Quarles, G.J.; Messing, G.L. Sintering and grain growth in SiO$_2$ doped Nd:YAG. *J. Eur. Ceram. Soc.* **2008**, *28*, 1527–1534. [CrossRef]
40. Marder, R.; Chaim, R.; Estournès, C. Grain growth stagnation in fully dense nanocrystalline Y$_2$O$_3$ by spark plasma sintering. *Mater. Sci. Eng. A* **2010**, *527*, 1577–1585. [CrossRef]
41. Chen, P.-L.; Chen, I.-W. Grain boundary mobility in Y$_2$O$_3$: Defect mechanism and dopant effects. *J. Am. Ceram. Soc.* **1996**, *79*, 1801–1809. [CrossRef]
42. Chaim, R.; Shlayer, A.; Estournes, C. Densification of nanocrystalline Y$_2$O$_3$ ceramic powder by spark plasma sintering. *J. Eur. Ceram. Soc.* **2009**, *29*, 91–98. [CrossRef]
43. Zhang, Y.; Navrotsky, A.; Li, H.; Li, L.; Davis, L.L.; Strachan, D.M. Energetics of dissolution of Gd$_2$O$_3$ and HfO$_2$ in sodium alumino-borosilicate glasses. *J. Non-Cryst. Solids* **2001**, *296*, 93–101. [CrossRef]
44. Bykov, Y.V.; Egorov, S.V.; Eremeev, A.G.; Kholoptsev, V.V.; Plotnikov, I.V.; Rybakov, K.I.; Sorokin, A.A. Sintering of Oxide Ceramics Under Rapid Microwave Heating. In *Processing, Properties and Design of Advanced Ceramics and Composites (Ceramic Transactions, vol. 259)*; Singh, G., Bhalla, A., Mahmoud, M.M., Castro, R.H.R., Bansal, N.P., Zhu, D., Singh, J.P., Wu, Y., Eds.; Wiley: Hoboken, NJ, USA, 2016; pp. 233–242.
45. De Gennes, P.G. On a relation between percolation theory and the elasticity of gels. *J. Phys. Lett.* **1976**, *37*, 1–2. [CrossRef]
46. Van Arkel, A.E.; Flood, E.A.; Bright, N.F.H. The electrical conductivity of molten oxides. *Can. J. Chem.* **1953**, *31*, 1009–1019. [CrossRef]

materials

MDPI

Article

Discussion on Local Spark Sintering of a Ceramic-Metal System in an SR-CT Experiment during Microwave Processing

Yongcun Li [1], Feng Xu [2,*], Xiaofang Hu [2], Bo Dong [2], Yunbo Luan [1] and Yu Xiao [2]

[1]   College of Mechanics, Taiyuan University of Technology, Taiyuan 030024, China;
      liyongcun@tyut.edu.cn (Y.L.); luanyunbo@tyut.edu.cn (Y.L.)
[2]   CAS Key Laboratory of Mechanical Behavior and Design of Materials, Department of Modern Mechanics,
      University of Science and Technology of China, Hefei 230026, China; huxf@ustc.edu.cn (X.H.);
      dongbo@mail.ustc.edu.cn (B.D.); xiaoyuxy@mail.ustc.edu.cn (Y.X.)
*   Correspondence: xufeng3@ustc.edu.cn; Tel.: +86-551-6360-0564

Academic Editor: Eugene A. Olevsky
Received: 15 January 2016; Accepted: 16 February 2016; Published: 26 February 2016

Abstract: In this paper, local spark sintering of a ceramic-metal system (SiO_2-Sn) during microwave processing was examinedby means of synchrotron-radiation-computed tomography technology. From the reconstructed 3-D and cross-section images of the specimen, adensification process was observed below the melting point of Sn, and then the specimen came into a rapid densification stage. These results may be due to the local spark sintering induced by the high-frequency alternating microwave electric fields. As the metallic particles Sn were introduced, the microstructure of "ceramic-metal" will lead to a non-uniform distribution and micro-focusing effect from electric fields in some regions (e.g., the neck). This will result in high-intensity electric fields and then induce rapid spark sintering within the micro-region. However, in the subsequent stage, the densification rate declined even when the specimen was not dense enough. The explanation for this is that as the liquid Sn permeated the gaps between SiO_2, the specimen became dense and the micro-focusing effect of electric fields decreased. This may result in the decrease or disappearance of spark sintering. These results will contribute to the understanding of microwave sintering mechanisms and the improvement of microwave processing methods.

Keywords: microwave sintering; microstructureevolution; metal; synchrotron radiation computed tomography

1. Introduction

Over the last decades, microwave sintering has been under constant development for the rapid preparation of high-performance powder materials, such as ceramics and ceramic matrix composites [1–5]. Recently, since the first full sintering of metal powders in a microwave field, many experiments on microwave sintering of various kinds of metals, including ceramic-metal materials, have been carried out. For example, E. Breval *et al.* [6] indicated that in the microwave sintering of WC-Co, there was very little WC particle growth and the specimen possessed six times more resistance against corrosion and a hardness 1–5 GPa higher than the conventional specimens.

Thus far, much work has been performed to study the mechanisms of microwave sintering. Many researchers attribute the advantages of microwave sintering to the effect induced byhigh-frequency alternating microwave fields, such as the enhancement of the diffusion coefficient [7,8], reduction of activation energy [9,10], micro-focusing effect [11] and the eddy current [12] for ceramics or metals. However, as for the metal-ceramic materials, due to the role of high-energy microwave fields, as well as the heterogeneity and non-uniform distribution of the mixed materials, there might besome special interaction mechanisms that are different from the microwave sinteringof pure ceramicand metal. For example, as the WC and Co mixed together, the heating efficiency of WC/Co in the magnetic-field strangely became lower than both the WC and Co [13]. These results indicate thattheinteraction mechanisms of themixed systemarenot just a simple superposition of the original mechanisms. These mechanisms may affect the microstructure and macro-performance of materials. This means it is quite necessary to explore the sintering mechanisms in the microwave sintering of metal-ceramic materials.

At present, the technology of TEM, SEM, hot-stage microscopes, *etc.* are usually adopted to study microwave sintering mechanisms [12,14]. These methods can be usedtocarry out the *in situ* analysis of the surface or slice information of a specimen. However, owing to the hightemperature, microwave radiation and the opacity of materials, it is difficult to realize theinternal and real-time microstructure evolution observation continuously during microwave processing. The synchrotron radiation computer tomography (SR-CT) technique is the latest non-destructive testing technology [15] based on the excellent synchrotron radiation light source (e.g., high intensity, strong penetrability and good coherence). It can achieve internal and real-time microstructure evolution observation of materials under extreme conditions (e.g., high temperature, intense radiation). By applying this technique, the surface and internal microstructure evolution during high-temperature microwave processing can be directly and continuously observed.

In this paper, the SR-CT technique was adopted toinvestigate the microstructure evolution of ceramic-metal system (silicon dioxide and tin, SiO_2-Sn) during microwave sintering. In theexperiment, 3-D and cross-section images of the same microstructureat differenttimes were obtained, and some typical sintering phenomena were clearly observed. Also, a densification processwas observed below the melting point of Sn, and then the specimen came into a rapid densification stage. Particle rotation and rearrangement was frequently observed. However, in the subsequent stage, the densification rate declined even when the specimen was not dense enough. The reason may be due to the decrease or disappearance ofspark sintering. As the liquid Sn permeated the gaps between particles, the specimen became dense and themicro-focusing effect of microwave electric fields decreased as a result. These results will contribute to the understanding of the rapid microwavesintering mechanisms of materials and the improvement of microwave processing methods.

2. Materials and methods

2.1. Materials

In our experiment, chemically pure SiO_2 (purity 99.9%, average diameter 150 μm) and Sn (purity 99.8%, average diameter 75 μm) powders were used. Before theexperiment, SiO_2-Sn powders (volume ratio of 1:10) were mixed uniformly in the anhydrous ethanol by the mechanical agitator for 4 h, then dried in the vacuum drying oven and loosely encapsulated into a closed quartz capillary (height: 10 mm, internal radius: 0.35 mm).

2.2. The SR-CT Experiment during Microwave Processing

The SR-CT experiment on microwave sinteringof SiO_2-Sn was carried out on the BL13W1 beam line at the Shanghai Synchrotron Radiation Facility (SSRF, Shanghai, China). In the experiment, the specimen was introduced into a specially designed microwave furnace (multimode cavity: 2.45 GHz, output power: 3 kW). The sintering temperature wasmeasured by a thermo tracer [type TH5104, range: 10–1500 °C, accuracy ± 1.0% (full scale)], and the typical temperature profile is shown in Figure 1.The temperature was compared with and calibrated by the thermocouple. Other goodresearch on calibration of temperature during the microwave sintering process is available elsewhere [16]. In the experiment, marking points (Cu particleswith a radius of about 20 μm were affixed on the capillary surface as marking points) were used to roughly track the same part of the sample, which usually contains several cross-sections. After finding the same part of the sample at different sintering times, one of the cross-sections images (gray images) within this part of the sample can be selected, and then the related algorithm can be employed to identify this cross-section with other cross-sections of the sample at different sintering times. Two cross-sections were considered the same cross-sections of this part of the sample when the correlation coefficient between these two figures reached the largest value.

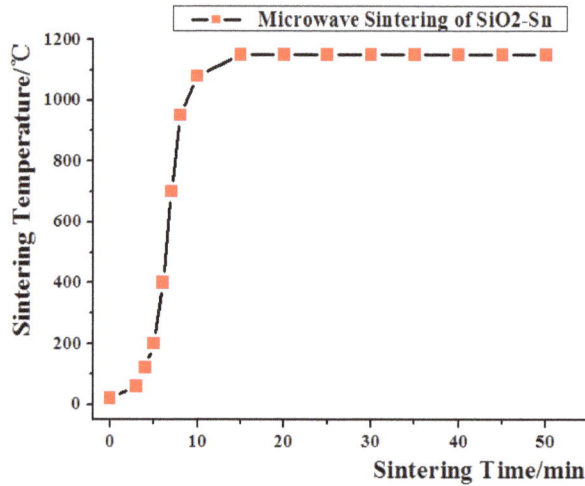

Figure 1. Typical temperature profile of SiO_2-Sn during microwave sintering.

3. Results and Discussion

Figure 2 shows the microstructure evolution of the same 3-D and cross-section of SiO_2-Sn atdifferent sintering times. Grayscale range is from 0 to 255; the closer to 255, the higher the relative density, which means that white represents particles and black represents pores. In Figure 2B, some typical particles were marked with numbers and letters (particles in blue color represent Sn, and red color represents SiO_2). From these images, the evolution of the same microstructurecan be tracked. For example, after microwave processing for some minutes, most of separated particles in Figure 2B(a) connected with each other and sinteringnecks formed in Figure 2B(b). The pores were interconnected with each other in Figure 2B(a), while in Figure 2B(d) the shape of pores changed and became closed and isolated. This means the specimen gradually changed from loose to dense. These phenomena were in accordance with the sintering theory and can also be observed insolid microwave sintering (e.g., Al-SiC, [17]). In addition, there were some special sintering phenomena. For example, the microstructure evolution can be well tracked during solid microwave sintering of Al-SiC [17]. However, in the microwave sintering of SiO_2-Sn, the rearrangement and rotation of particles were more frequent, and it wasvery hard to track the same microstructure during sintering. As we know, in microwave sintering the heat is derived from the direct interaction between microwave and material. This means the heating characteristic within the material is inevitably related to the microstructure and itsmaterial evolution. In order to investigate microwave sintering mechanisms, quantitative analysis of microstructure evolution parameters is an effective and important method.

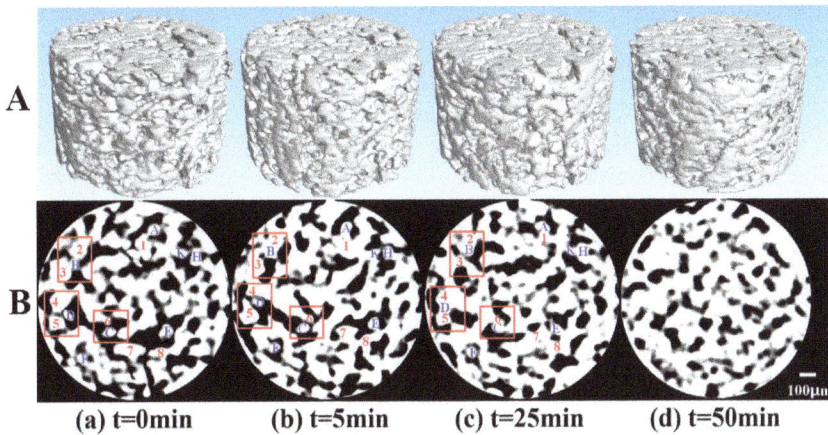

Figure 2. 3-Dand cross-section images of the evolution of the same the same cross-sectionat different sintering times.

In this section, the statistics of the porosity of specimen in the differenttimes were carried out. The porosity represents the local bulk percentage of pore in the specimen. It is the averageof several similar cross-sections. The results are shown in Figure 3A. In the sintering theory, the densification parameter $\alpha$is usually adopted to analyze the densification process of sintering. It can be described as follows,

$$\alpha = \frac{\rho - \rho_0}{\rho_1 - \rho_0} \times 100\% \tag{1}$$

Here, ρ represents true density, and ρ_0 and ρ_1 representthe density of the green body and theoretical density, respectively. This means the densification parameter α can be described as

$$\alpha = \frac{\rho/\rho_1 - \rho_0/\rho_1}{1 - \rho_0/\rho_1} \times 100\% = \frac{(1-\beta) - (1-\beta_0)}{1 - (1-\beta_0)} \times 100\% = \frac{\beta_0 - \beta}{\beta_0} \times 100\% \tag{2}$$

Here β represents true porosity, and β_0 represents the porosity of the green body. Inorder to study the densification process of SiO$_2$-Sn during microwave sintering, the relationship between α and sintering time is shown in Figure 3B. In Section 3.1, the densification process and sintering mechanisms ofthe microstructure evolutionof the specimen will be discussed.

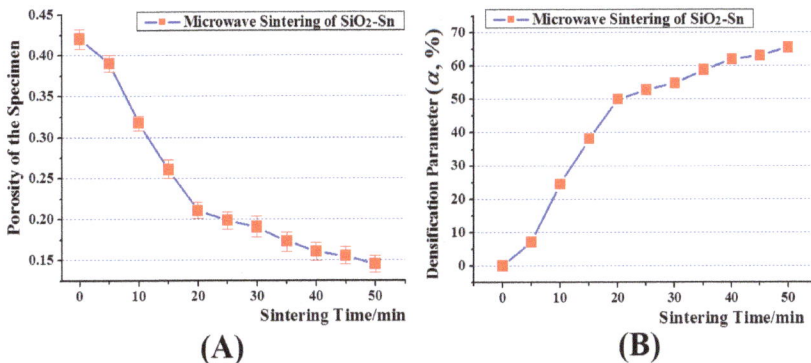

Figure 3. (**A**) The relationship between porosity and sintering time; and (**B**) the relationship between densification parameter and sintering time.

3.1. Densification below the Melting Point of Sn Induced by Microwave E-Fields

Figure 3B shows the change in α at different sintering times. It can be seen that the densification rate was very largefrom the 5th to 20th minute, and declined gradually from the 20th to 50th minute. Moreover, at the beginning, the value of α began to rise from 0 to 7.08 in the first 5 minute. However, the highest sintering temperature during this period was only 200 °C, which was below the melting point of Sn (231 °C). Although the sintering would occur between particles of Sn at this temperature, the volume ratio of Sn is only 1/11, and most Sn particles were separated by SiO_2 particles. How did the decline of porosity occur?

Figure 3 shows that the main densification began to occurin the first 5 minute. The corresponding microstructure evolution during this period is shown in Figure 2B(a,b). From these images, the microstructure evolution can be clearly observed. For example, particle B was separated from particles 2 and 3 at the beginning. However, until the 5th minute, particle B was sintered together with particles 2 and 3. The local amplification images of these particles are shown in Figure 4a similar microstructure evolution phenomenon was also observed on the other cross-section of the sample. This will inevitably lead to the densification of specimen.This phenomenon may be induced by the spark sintering. The explanation is that, as Figure 4b shows, the sintering has happened between particles B, 2 and 3. This rapid sintering in such a short time and below the melting point of Sn indicates that there was a lot of heat deposition between particles induced by microwave fields, which then led to the high temperature within the local connection regions between particles. This highly energy deposition and high temperature is most likely due to the spark sintering caused by microwave electric fields.

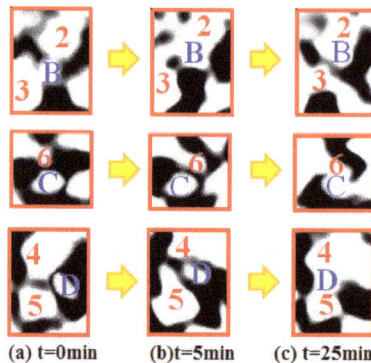

(a) t=0min   (b)t=5min   (c) t=25min

Figure 4. The local amplification images of particles B, C and D in Figure 2B.

As we know, spark sintering is a rapid sintering method viaelectric spark between particles induced by current. Yet in microwave sintering, the microwave energy is the direct energy source for the sintering process. This meansspark sintering likely must be closely related to the intensity distribution of microwave E-fields. In order to further study the reasons for spark sintering, it is necessary to research the intensity distribution of microwave E-fields within the specimen.

To study the intensity distribution of microwave E-fields, the TE101 mode resonator (109.2 mm × 54.6 mm × 74 mm) was used. With this resonator, the distribution of microwave electric fieldwas analysed by using the HFSS finite element analysis software. The TE101 mode resonator is a single-mode microwave resonator. It sets a 1 and 1 standing wave cycle of the electric fields in the direction of the x and z axes, respectively, and sets the electric field to be uniform and parallel to the y axis. In this way, an investigation into the propagation characteristics of microwave E-fields in a certain direction can be performed by using this resonator. To compare the obtained results with those obtained experimentally, the microwave frequency is set to 2.45 GHz, and the solution type is set to driven modal. The ceramic particle of SiO_2 with a diameter of 150 µm and metallic particle Sn with a diameter of 75 μm is selected as the material model, which is identical to that used experimentally. Using this model, the E-field distribution between two of the connected particles was examined. Appropriate relative permittivity and relative magnetic permeability of 9.8 and 1.0, respectively, are assigned to the ceramic particles of SiO_2. The appropriate relative permittivity and relative magnetic permeability of 1.0 and 1.0, respectively, are assigned to the metallic particle of Sn. Note that other values can be chosen for these two parameters if other materials are simulated. To investigate the behaviour of the microwave E-fields independently, the particles were placed in the region where the E-fields are at a maximum and where the H-fields are almost zero.

In the TE101 rectangular single-mode cavity, an operating frequency of 2.45 GHz will produce a wavelength around 122.5 mm, and the half wavelength of the standing wave are 109.2mm and 74mm in the direction of the x and z axes, respectively. Comparing the size of the particles (diameter 0.075 mm) with the half wavelength of the standing wave, the microwave E-fields in the sample can be considered to be approximately homogeneous prior to the introduction of the particles. Figure 5 shows the intensity distribution of microwave E-fields when the particle Sn is located within the sintering neck of SiO_2, which is similar to the case of particles B, C and D located within the sintering neck of SiO_2 particles 2 to 6.

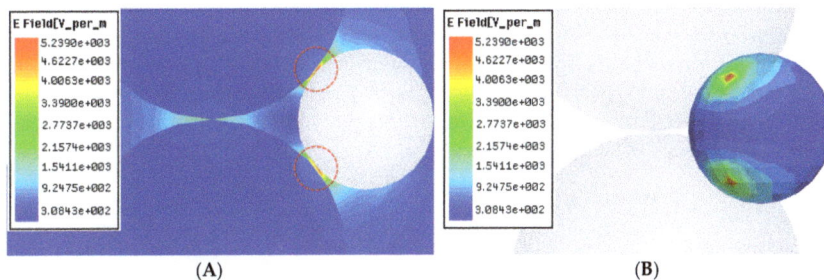

Figure 5. The intensity distribution of microwave E-fields (**A**) within the contacted particles of SiO_2-Sn; and (**B**) on the surface of particle Sn.

As can be seen, the E-fields within the micro-region of sintering neck between SiO_2 are much larger than the average fields, which suggests that the E-fields are focused in this micro-region. However, it can also be seen that the E-fields within the micro-region of the sintering neck between SiO_2-Sn are much larger than even the E-fields between SiO_2, which indicates that the focusing effect of E-fields is much more significantin this local micro-region.In the simulation, we assume that the intensity of the applied field is 3.30×10^2 v/m. Then it can be determined that the peak E-field between SiO_2-Sn within the neck region is 5.24×10^3 v/m, which is about 15.88 times larger than the applied field. In the actual sintering, the intensity of the applied field will be very high, so the peak E-field within the neck region between SiO_2-Sn will be much higher. For example, if the average E-field during microwave sintering is 1 kV/cm (e.g., ZnO [11]), the peak E-fields can be as high as 15.88 kV/cm. Because the square amplitude of the electric field E^2 is directly related to the power density of microwaves [11], and thus the heating rate, the peak microwave energy is approximately 250 times greater than the applied microwave energy at the beginning of the sintering process, which may be large enough to cause microscopic ionization at atmospheric pressure and lead to spark sintering [11,18,19]. Therefore, although the applied field is not very high and the overall sample temperature is below the melting point of Sn, the peak E-fields and temperature within the micro-region between SiO_2-Sn will be high enough and induce local spark sintering. As a result, mass diffusion will be increased and the sintering process will be accelerated as well. This factor may be an important mechanism for rapid preparation involving microwave sintering.

3.2. Rapid Densification of the Specimen in the Middle Sintering Period

As the sintering process wenton, the sintering temperature increasedrapidly (e.g., by the 10th minute the temperature reached 1080 °C)and thedensification rate became quite large. As shown in Figure 2B, particles C and D were still separated fromparticles 6 and 4, 5 at the 5th minute. However, in the 25th minute these particles were sintered together.Thelocal amplification images of particles B, C and D in Figure 2B are shown in Figure 4.

This phenomenon may also be due to thespark sintering, because there was initially a distance between particles C and 6, as well as between particles D and 4, 5. As the sintering process wenton, the particles Sn entered aliquid phase and the sample started to shrink. This may make these particles contact each other. This means spark sintering may occur, induced by the micro-focusing effect of microwave E-fields between these particles. This is the same as the case that happened between particles B and 2, 3 in the first 5 minutes. Besides, as the sintering neck between particles grew larger, there was still a focusing effect between particles. Figure 6a shows the distribution of microwave E-fields on the particle surface when the sinteringneck is 0.12 times the average of the radius of Sn and SiO_2 [there was still a pore between the particles of SiO_2-Sn and the sintering state is between the sintering states of Figure 4b,c]. From Figure 6a, it can be seen that there is still a focusing effect within the micro-region of the sintering neck between SiO_2-Sn. Here, the peak E-field between SiO_2-Sn within the neck region is 3.74×10^3 v/m, which is about 11.33 times larger than the applied field (3.30×10^2 v/m).This means that in the actual sintering, the intensity of the peak electric field within the neck region between SiO_2-Sn will be 11.33 times larger than the applied fields, and the peak E^2 field is approximately 128 times greater than the applied E^2 field, which may also cause microscopic ionization and lead to spark sintering.

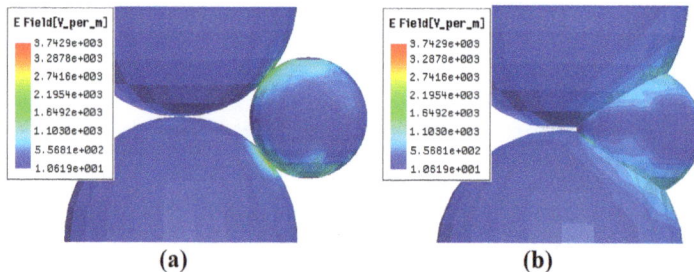

(a)   (b)

Figure 6. The intensity distribution of microwave E-fields in the particles of SiO_2-Sn at different sintering stages (**a**) the sintering neck is 0.12 times of the average of the radius of Sn and SiO_2; and (**b**) the distance between the centers of Sn and SiO_2 is 0.75 times of the sum of the radius of Sn and SiO_2.

3.3. Decline of Densification Rate when the Specimen Was not Dense Enough

Section 3.1 has discussed the rapid densification process of SiO_2-Sn during microwave sintering from the 5th to 25th minute. However, from the 20th minute, the densification rate declined a lot (as shown in Figure 3B). This phenomenon also happens in the conventional liquid sintering, and usually happens when the densification parameter reaches about 75~80 [20]. However, in this study the densification parameter of the specimen during this period was only 50~65. This means that there were still considerable pores and the specimen was not dense enough, like the image shown in Figure 2B(d). There are some reasons for this phenomenon: Firstly,the highest sintering temperature is about 1150°C, which is a little lower than the usually sintering temperature of SiO_2 (>1200 °C). This means thatin the laterstage, the solid sintering rate between SiO_2 particles would be blocked because of the lower temperature. Secondly, due to the small volume fraction of Sn (10%), as the specimen became densethe distance between particles decreased at the same time, and the viscous flow of Sn and particle rearrangement would decline gradually.Besides, the decrease of the distance between particles will make the friction between particles increases, which will inhibit the further densification ofthe specimen. Thirdly, as the specimen became dense, the liquid phase of Sn penetrated the gap between particles. This will produce a change in microstructure configuration and may result in the reduction or disappearance of spark sintering. Figure 6b shows the distribution of microwave E-fields on the particle surface of SiO_2-Sn (the distance between the centers of Sn-SiO_2 is 0.75 times of the sum of the radius of Sn and SiO_2). This sintering state is between the sintering states of Figure 4b,c. From this figure, it can be seen that the peak E-field between SiO_2 and Sn within the neck region is 1.10×10^3 v/m. This means that the intensity of peak E-fields within the neck region between SiO_2 and Sn was only 3.33 times that of the applied fields, and cannot be large enough to cause microscopic ionization at atmospheric pressure and lead to spark sintering. This meansthe driving force for the mass diffusion decreased and the sintering process including the densification process declined as a result. These may be the reasons why the densification rate declined when the specimen was not dense enough.

4. Conclusions

An *in situ* investigation on local spark sintering of ceramic-metal system (SiO_2-Sn) during microwave processing was carried out using synchrotron radiation computed tomography technology (SR-CT).The following conclusions can be drawn:

In the experiment, except for the normal sintering phenomenon, a densification process was observed below the melting point of Sn, and then the specimen came into a rapid densification stage. This result may be due to the spark sintering induced by the high-frequency alternating microwave electric fields.

In the ceramic-metal system (Sn-SiO_2), as the metallic particles were introduced, the microstructure of "ceramic-metal" will lead to a non-uniform distribution and micro-focusing effectofelectric fields in some regions (e.g., the neck). This will result in high intensity electric fields and may induce the local rapid spark sintering within micro-region.

In the later sintering stage, as the liquid Sn permeated the gaps between SiO_2, the specimen became dense and the micro-focusing effect ofelectric fields decreased. This may result in the decrease or disappearance of spark sintering, so the densification rate declined even when the specimen was not dense enough.

These mechanisms may be the explanation for the microstructure evolution process during microwave sintering of SiO_2-Sn, and may contribute to the understanding of microwave sintering mechanisms and the improvement of microwave processing methods.

Acknowledgments: The authors Feng Xu and Yongcun Li are co-first authors, and they contributed equally to the work.This research was supported by the National Natural Science Foundation of China (Nos. 11272305, 11402160, 11472265, 11172290, 21501129), the National Basic Research Program of China (973 Program, No.2012CB937504)and Anhui Provincial Natural Science Foundation (No. 1508085MA17).

Author Contributions: Feng Xu and Yongcun Li designed the experiments. Bo Dong and Yu Xiao performed the SR-CT experiments; all the authors contributed to the data analysis and discussion; Yongcun Li wrote the paper.

Conflicts of Interest: The authors declare no conflict of interest.

References

1. Yadoji, P.; Peelamedu, R.; Agrawal, D.; Roy, R. Microwave sintering of Ni–Zn ferrites: Comparison with conventional sintering. *Mat. Sci. Eng. B* **2003**, *98*, 269–278. [CrossRef]
2. Agrawal, D. Latest global developments in microwave materials processing. *Mater. Res. Innov.* **2010**, *14*, 3–8. [CrossRef]
3. Clark, D.E.; Folz, D.C.; West, J.K. Processing materials with microwave energy. *Mat. Sci. Eng. A* **2000**, *287*, 153–158. [CrossRef]
4. Morteza, O.; Omid, M. Microwave versus conventional sintering: A review of fundamentals, advantages and applications. *J. Alloy. Compd.* **2010**, *494*, 175–189.

5. Cheng, J.; Agrawal, D.; Zhang, Y.; Roy, R. Microwave sintering of transparent alumina. *Mater. Lett.* **2002**, *56*, 587–592. [CrossRef]

6. Breval, E.; Cheng, J.P.; Agrawal, D.K.; Gigl, P.; Dennisb, M.; Roya, R.; Papworth, A.J. Comparison between microwave and conventional sintering of WC/Co composites. *Mat. Sci. Eng. A* **2005**, *391*, 285–295. [CrossRef]

7. Wilson, B.A.; Lee, K.Y.; Case, E.D. Diffusive crack-healing behavior in polycrystalline alumina: A comparison between microwave annealing and conventional annealing. *Mater. Res. Bull.* **1997**, *32*, 1607–1616. [CrossRef]

8. Janney, M.A.; Kimrey, H.D.; Schmidt, M.A.; Kiggans, J.O. Grain growth in microwave-annealed alumina. *J. Am. Ceram. Soc.* **1991**, *74*, 1675–1681. [CrossRef]

9. Janney, M.A.; Kimrey, H.D.; Allen, W.R.; Kiggans, J.O. Enhanced diffusion in sapphire during microwave heating. *J. Mater. Sci.* **1997**, *32*, 1347–1355. [CrossRef]

10. Demirskyi, D.; Agrawal, D.; Ragulya, A. Neck growth kinetics during microwave sintering of nickel powder. *J. Alloy. Compd.* **2011**, *509*, 1790–1795. [CrossRef]

11. Birnboim, A.; Calame, J.P.; Carmel, Y. Microfocusing and polarization effects in spherical neck ceramic microstructures during microwave processing. *J. Appl. Phys.* **1999**, *85*, 478–482. [CrossRef]

12. Ma, J.; Diehl, J.F.; Johnson, E.J.; Martin, K.R.; Miskovsky, N.M.; Smith, C.T.; Weisel, G.J.; Weiss, B.L.; Zimmerman, D.T. Systematic study of microwave absorption, heating, and microstructure evolution of porous copper powder metal compacts. *J. Appl. Phys.* **2007**, *101*, 074906–1–8. [CrossRef]

13. Cheng, J.P.; Roy, R.; Agrawal, D. Radically different effects on materials by separated microwave. *Mater. Res. Innov.* **2002**, *5*, 170–177. [CrossRef]

14. Subhadip, B.; Susmita, B.; Bandyopadhyay, A. Densification study and mechanical properties of microwave-sintered mullite and mullite–zirconia composites. *J. Am. Ceram. Soc.* **2011**, *94*, 32–41.

15. Li, X.; Hu, X.F. Synchrotron radiation tomography for reconstruction of layer structures and internal damage of composite material. *Chin. J. Lasers B* **1999**, *6*, 503–508.

16. Zuo, F.; Saunier, S.; Marinel, S.; Chanin-Lambert, P.; Peillon, N.; Goeuriot, D. Investigation of the mechanism(s) controlling microwave sintering of α-alumina: Influence of the powder parameters on the grain growth, thermodynamics and densification kinetics. *J. Eur. Ceram. Soc.* **2015**, *35*, 959–970. [CrossRef]

17. Li, Y.C.; Xu, F.; Hu, X.F.; Kang, D.; Xiao, T.P.; Wu, X.P. In situ investigation on the mixed-interaction mechanisms in the metal–ceramic system's microwave sintering. *Acta Mater.* **2014**, *66*, 293–301. [CrossRef]

18. Su, H.; Lynn, J.D. Sintering of Alumina in Microwave-Induced Oxygen Plasma. *J. Am. Ceram. Soc.* **1996**, *79*, 3199–3210. [CrossRef]

19. Nouari, S.; Zafar, I.; Abdullah, K.; Hakeem, A.S.; Nasser, A.A.; Tahar, L.; Amro, A.Q.; Kirchner, R. Spark plasma sintering of metals and metal matrix nanocomposites: A review. *J. Nanomater.* **2013**, *2012*, 4873–4881.

20. Kingery, W.D. Densification during sintering in the presence of a liquid phase. II.Experimental. *J. Appl. Phys.* **1959**, *30*, 307–310. [CrossRef]

materials

MDPI

Article

Microstructure Investigation of 13Cr-2Mo ODS Steel Components Obtained by High Voltage Electric Discharge Compaction Technique

Igor Bogachev [1,*], Artem Yudin [1], Evgeniy Grigoryev [1], Ivan Chernov [1], Maxim Staltsov [1], Oleg Khasanov [2] and Eugene Olevsky [1,2,3]

[1] Material Science Department, Moscow Engineering Physics University, Moscow 115409, Russia; halfdeath0@gmail.com (A.Y.); eggrigoryev@mephi.ru (E.G.); i_chernov@mail.ru (I.C.); msstaltsov@mephi.ru (M.S.); eolevsky@mail.sdsu.edu (E.O.)
[2] Department of Nanomaterials and Nanotechnologies, National Research Tomsk Polytechnic University, Tomsk 634050, Russia; khasanov@tpu.ru
[3] Department of Mechanical Engineering, San Diego State University, San Diego, CA 92182, USA
* Correspondence: iabogachev@mephi.ru; Tel.: +7-495-788-5699

Academic Editor: Jai-Sung Lee
Received: 9 October 2015; Accepted: 27 October 2015; Published: 2 November 2015

Abstract: Refractory oxide dispersion strengthened 13Cr-2Mo steel powder was successfully consolidated to near theoretical density using high voltage electric discharge compaction. Cylindrical samples with relative density from 90% to 97% and dimensions of 10 mm in diameter and 10–15 mm in height were obtained. Consolidation conditions such as pressure and voltage were varied in some ranges to determine the optimal compaction regime. Three different concentrations of yttria were used to identify its effect on the properties of the samples. It is shown that the utilized ultra-rapid consolidation process in combination with high transmitted energy allows obtaining high density compacts, retaining the initial structure with minimal grain growth. The experimental results indicate some heterogeneity of the structure which may occur in the external layers of the tested samples due to various thermal and electromagnetic in-processing effects. The choice of the optimal parameters of the consolidation enables obtaining samples of acceptable quality.

Keywords: field-assisted; powder consolidation; steel; oxide dispersion strengthening

1. Introduction

Refractory materials, which are used in the fast reactor active zone, should meet a number of special requirements. However, many of the materials currently utilized in fuel claddings and other components of the reactor active zones do not meet the levels of the required mechanical properties, radiation, and thermal resistance, especially for the new projects on high-power nuclear reactors. They should have high thermal resistance at elevated temperatures, low thermal creep, low swelling, and low properties of degradation under irradiation. Also, they should possess suitable mechanical properties to withstand the pressure of a heat transfer agent and fuel. Corrosion resistance is also necessary. Currently, in many cases, special austenitic steels such as 16Cr-15Ni-2Mo steel are used as the material for cladding tubes for fast nuclear reactor active zones. These steels (with FCC crystal lattice) have suitable mechanical properties and radiation resistance at elevated temperatures and high neutron fluence [1,2]. However, in the next generation of fast reactors, for more energy efficiency the operating temperature will be higher, and the resulting neutron fluence will be higher due to the increase of the fuel campaign [2–4]. Alternatively, ferritic/martensitic steels with BCC lattice could be used for these applications because of their higher radiation resistance and, in particular, because of low swelling in comparison to austenitic steels. Yet these steels have quite high thermal creep at elevated temperatures which limits their use in the active zone [2–4]. To improve this parameter, oxide dispersion strengthening (ODS) of the matrix material by the refractory hard micro- or nanoscale particles is used. This approach enables obtaining thermal stability of the matrix steel, inhibiting dislocation movement and, as a result, decreases thermal creep [5–12]. Thereby, ODS steels are a prospective material for nuclear applications. The development of efficient production routes and obtaining suitable properties of the produced ODS materials are important modern problems whose solutions should enable the use of these steels in nuclear reactors. In most cases, ODS steels are produced via a powder processing route [5–7]; therefore, the mechanical properties of the fabricated ODS steel components become a major controlling factor.

The present work explores the possibility of a new method for the consolidation of ODS steel components by employing the high voltage electric discharge compaction technique. It allows obtaining homogeneous distribution of material components, controlling mechanical and physical properties of the final product at the stage of manufacturing, and achieving different final shapes of the product without any intermediate treatment [13–15]. It should be noted that electromagnetic field-assisted powder consolidation, such as spark plasma sintering, microwave sintering, *etc.*, has great advantages for the processing of such materials and, in particular, of oxide dispersion-strengthened steels. The high voltage electric discharge compaction (HVEDC) technology occupies a special place among the above-mentioned techniques. It has unique operation parameters in comparison with other electromagnetic field-assisted techniques [16–18].

High voltage electric discharge compaction technology has a number of advantages in comparison with conventional powder processing techniques such as hot extrusion, hot and cold isostatic pressing, and others. This method involves the uniaxial pressing of the powder in a non-conducting matrix with discharging a powerful pulse of electric current through the specimen. Electric energy of a set value is stored in a block of high-voltage capacitors and pressure is controlled by a pneumatic press.

Such an ultra-rapid process in combination with high transmitted energy allows obtaining high density compacts, saving the initial structure with minimal grain growth, and avoiding thermally activated phase transformations. All these parameters are important for the ODS steels compaction because there is a need to obtain homogeneous distributions of the hard refractory oxide particles inside the matrix steel powder to prevent their agglomeration and grain growth for better mechanical properties. Also, this manufacturing technology has the ability of final product net-shaping. Using various shapes of punches and dies, the freeform products can be manufactured.

2. Materials and Methods

The initial base material for the conducted investigations was ferritic/martensitic 13Cr-2Mo special reactor steel, with the chemical composition shown in Table 1.

Table 1. Chemical composition of the investigated steel powder.

C	Si	Mn	Cr	Ni	Mo	Nb	V
0.10–0.15	<0.6	<0.6	12.0–14.0	<0.3	1.2–1.8	0.25–0.55	0.1–0.3

This material in the form of flakes with the average size of 2–3 mm in length and 200–300 μm in thickness was obtained by casting the melt onto a rapidly rotated cooled massive disk. These flakes were pre-milled in a planetary ball mill (MTI LCC, Richmond, CA, USA) for 2 h in air to the state of a powder with particle sizes of up to 400 μm. This powder was mechanically mixed with different concentrations of commercial nanoscale Y_2O_3 powder ("Advanced Powder Technologies" LCC, Tomsk, Russia) with the average particle size of about 50 nm (Figure 1) and then was mechanically alloyed in a high-energy planetary ball mill (Fritsch GmbH, Idar-Oberstein, Germany) for 30 h in argon atmosphere.

During mechanical alloying the milling parameters such as rotation speed, time, and operation/standby intervals were varied in some range to obtain the most homogeneous powder.

It was determined that the rotation speed and the working period of the mill may cause significant heating and agglomeration of the powder. Thus, these parameters were optimized to prevent such negative implications. On the other hand, the influence of the milling time on the particle size was investigated. The average particle size of the powder starts to reduce significantly at the beginning of the process and further mechanical alloying does not allow any significant reduction of the average particle size.

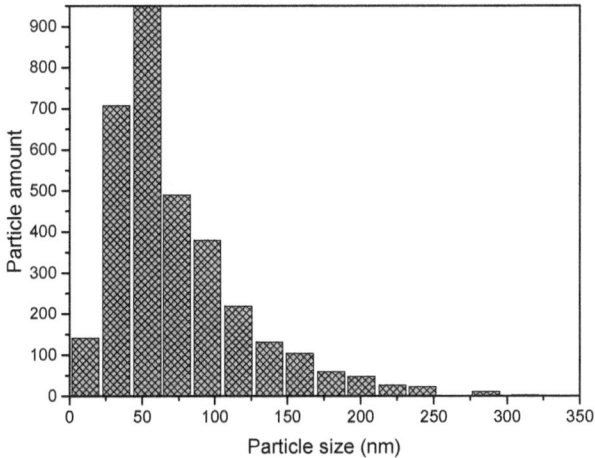

Figure 1. Particle size distribution of nanoscale Y_2O_3 powder.

More details on the influence of the milling time on the particle size distribution in this powder system are provided in Ref. [19]. As a result, the milling time of 30 h at the rotation speed of 200 rpm and a 2 h/2 h operation/standby cycle were chosen as optimal parameters for mechanical alloying. The chemical composition of the obtained powders indicates that there are no significant changes of the yttria concentration or an appearance of any impurities during mechanical alloying (Table 2).

Table 2. Chemical composition of the powder after mechanical alloying with 0.3 wt% of yttria.

Element	Fe	Si	Mn	Cr	Ni	Mo	Nb	V	Y
Concentration, wt%	83.11	0.058	0.43	12.9	0.11	1.4	0.20	0.29	0.31

To obtain the data on the effect of the yttria concentration on the densification and other properties of the powders, two different concentrations of Y_2O_3 were added to the powder batches at the stage of mechanical alloying of 0.3 and 0.7 wt%, and the powder without yttria content was also used in the comparative analysis.

The particle size analysis (Fritsch GmbH, Idar-Oberstein, Germany) shows that all powder batches with different yttria content have similar size distributions (Figure 2). The average particle size for all the batches was about 50 μm.

The high voltage electric discharge compaction was provided by the Impulse-BM device ("Potok" LCC, Rostov-on-Don, Russia) (Figure 3a) which enables an electric discharge through the specimen with a voltage of up to 6 kV and a pressure of up to 10 atm. This equipment allows the application of the voltage of several kilovolts and the pressure of up to several atmospheres to a powder specimen. Depending on these parameters, a density of the electric current in the powder can be achieved up to 500 kA/cm². The shape of the passing electric pulse is shown in Figure 3b and the time of the impact on the powder is up to several hundred milliseconds.

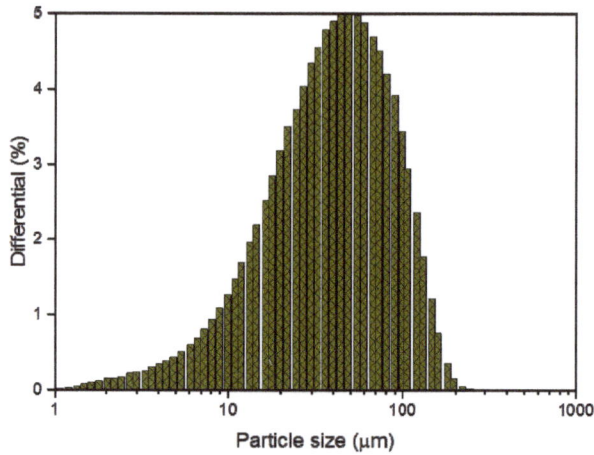

Figure 2. Typical particle size distribution of the obtained ODS steel powders.

The consolidation was conducted for all the specimens using similar parameters. The powder was loaded in a non-conductive ceramic die with a rugged metal collar and Mo conductive electrodes were used as punches. The consolidation was held in air atmosphere and the initial mass of the powder filling was 5 g. The initial green density was about 55%. Three different pressure values, 170, 200, and 270 MPa, and five voltage values, 1.5, 2.0, 4.0, 4.2, and 4.4 kV, were used in the experiments. The obtained cylindrical samples were 10 mm in diameter and 10 to 15 mm in height.

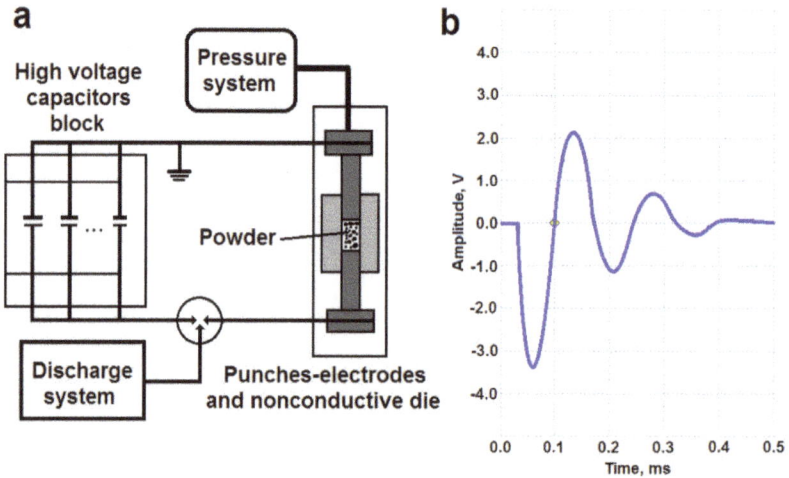

Figure 3. High voltage compaction principle scheme (**a**) and shape of the electric pulse (**b**).

Consolidated samples were mechanically polished and etched by 5 wt% nitric acid-alcohol mixture. The microstructure of the samples was investigated by means of optical microscopy. Microhardness data were obtained using a microhardness tester with 100 g loading and 15 s exposure. To obtain the data on the spatial density distribution inside the specimens' volume, the processed samples were cut in layers in the axial direction. The final density was measured using three techniques—geometrically, with hydrostatic weighing, and with helium pycnometry. Final density for all the samples was within the range from 90% to 97.5%.

3. Results and Discussion

Consolidation conditions and properties of the samples are shown in Table 3. It should be noted that the consolidation regime for two of the samples (B4 and B7) was unstable, leading to the knockout of the powder though the gap between the punches and the die. The weight of the samples does not match the weight of the powder filling due to the surface roughness of the punches and the die whereby a part of the powder can spill out of the die.

It should be noted that HVEDC is a fast and a metastable process associated with high heating rates, with cooling rates, and with changes of electrical conductivity of the processed medium, and with other factors. In this regard, unstable regimes of consolidation may occur as indicated by a rapid increase of the conductivity of the consolidated medium, an avalanche temperature increase, liquid phase formation, knockout of the powder from the die, and by other effects. These facts should be taken into account when interpreting the experimental results since they significantly affect the structural and phase state of the processed material and its physical properties.

Table 3. Conditions of the compaction and properties of the samples.

Sample	Y_2O_3 Content, wt%	m_f *, g	m_s *, g	P, MPa	U, kV	Mode	$\rho_{rel.}$, %
B5	0.0	5.00	4.03	270	1.5	Stable	92.18
B7	0.0	5.00	3.48	200	4.4	Unstable	95.83
B8	0.0	5.00	4.32	200	4.0	Stable	96.09
B3	0.3	5.00	4.27	200	4.2	Stable	97.48
B4	0.3	5.00	4.50	270	4.0	Unstable	96.32
B6	0.3	5.00	4.72	270	2.0	Stable	97.04
B9	0.7	5.00	4.66	170	4.0	Stable	90.30
B10	0.7	5.00	4.83	170	4.2	Stable	93.33
B11	0.7	5.00	4.67	170	4.4	Stable	93.10

Notes: * m_f—Mass of the powder filling; m_s—Mass of the sintered sample.

First, the samples' relative density dependence on the density of the electric current was determined (Figure 4). This approach clarifies how the transmitted electric power affects the compaction of the powder that allows the selection of optimal consolidation conditions. Since the electric current density depends on the resistivity of the powder at each moment of time, and it in turn depends on the pressure, the data was grouped according to the applied pressure. It is shown that the increasing pressure, in general, leads to the increasing final density. The samples, which were consolidated at 170 MPa pressure, have lower values of final density in comparison with the ones obtained at 200 or 270 MPa pressure. Also, the final density dependence on voltage was determined. An increase of the voltage values from 4.0 to 4.2 kV leads to the increasing relative density, but an insignificant decrease of the density was determined for the samples consolidated at 4.4 kV in comparison with the ones consolidated at 4.2 kV. This is valid for the whole range of used pressures. This effect may be associated with the occurrence of the local inhomogeneity of the electric current in the interparticle contacts due to which certain areas of the powder volume may include quite significant porosity. At the same time, samples that were consolidated at low values of the electric current density demonstrate quite high values of final density. This means that it is possible to obtain good-quality dense samples using quite low pressure and high voltage values and, at the same time, using high pressure but low voltage. Therefore, it should be noted that in the case of operating at low pressures, the conductivity of the powder composition may be insufficient for normal passing of the electric current. This may lead to an unstable consolidation process due to the high resistivity of the interparticle contacts resulting in the overheating of these areas, in the emergence of the electric breakdown between particles, and in the knockout of the powder. On the other hand, operating at high pressure values leads to better interparticle contacts and decreasing resistivity values of these contacts. This, in turn, leads to the lower energy release and heating of the powder which affects the sintering consolidation mechanism. It is clear that, using high pressures, it is possible to compact samples to high density values only by cold plastic deformation mechanisms, but strength and hardness of the produced samples may be insufficient. Also, quite strong die and punches are needed to realize such a process. Therefore, it is desirable to combine both mentioned mechanisms to obtain an acceptable product. As a result, 200 MPa and 4.2 kV consolidation conditions were chosen as optimal for the compaction of the ODS steel powder.

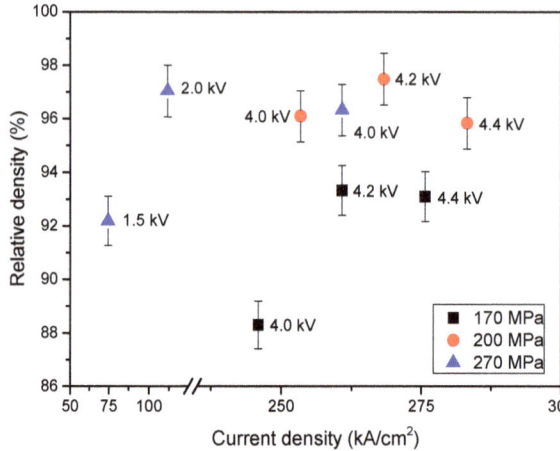

Figure 4. Samples' relative density dependence on the density of the electric current.

The next step of the investigation was to determine the density and microhardness distribution in the volume of the samples. To obtain the above-mentioned data, the processed samples were measured using a microhardness tester (FM-800, Future-Tech, Kawasaki, Japan; 100 g loading for 10 s) in axial and radial directions with 50 μm steps between each measurement point. Measured values had quite significant scatter, so the final data were presented in the form of areas. Then the samples were cut in the axial direction into several layers to determine the spatial density distribution. Figure 5 shows the axial microhardness distribution of the samples without the yttria addition. Sample B5 (1.5 kV, 270 MPa) demonstrates quite uniform microhardness with insignificant decreasing at the edges and the average value of 525 HV. Sample B8 was consolidated using a higher voltage of 4.0 kV and a pressure of 200 MPa. The average microhardness of this sample is about 650 HV, but the edges of the sample have a higher microhardness in comparison to its center. This effect could be caused by non-uniform consolidation of the powder. During electric discharge the powder areas located closer to the punches can densify more intensively due to higher values of plastic deformation in comparison with the central part [20]. Another reason for this effect may be associated with high heating and cooling rates of these areas. After the electric discharge the punches rapidly cool the powder in these areas due to their high thermal conductivity, and a tempered structure may be formed. The microhardness values of approximately 775 and 675 HV at the edges of the sample correspond to the ones of the tempered structure. In the case of sample B5, the tempering had not occurred due to an insufficient discharge power. The consolidation conditions of sample B7 were unstable, which greatly affected the variation of the microhardness values and which is also observed in the radial microhardness distribution (Figure 6). It is clear that an unstable consolidation may cause a significant inhomogeneity of the structure. Other samples show similar microhardness distributions. Increasing hardness of the edges of the samples in the radial direction in this case may be present due to two factors. One of them is the emergence of the skin-effect in the surface layers of the samples due to the high-frequency discharge with the frequency of 6–7 kHz. This effect causes an additional heating of these areas and intensive plastic deformation. Another reason is the forming of the tempered structure due to high cooling rates in comparison with those of the central areas. In general, the average microhardness values correlate with the density of the samples: B5 has 92% relative density and B8 has 96%.

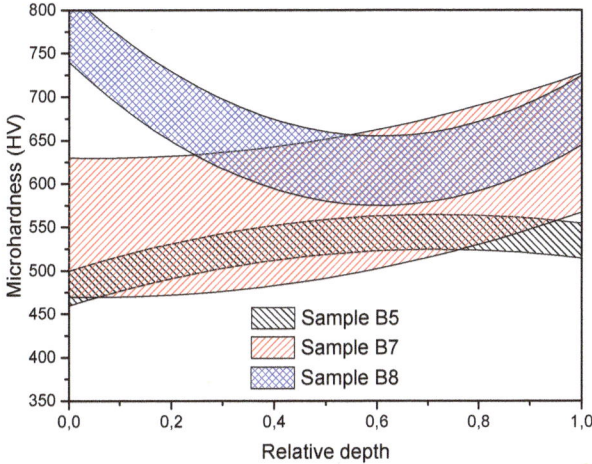

Figure 5. Axial microhardness distribution along samples' depth without yttria addition.

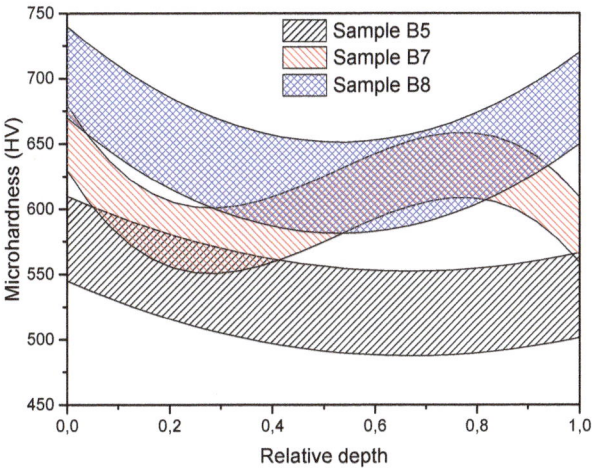

Figure 6. Radial microhardness distribution along samples' depth without yttria addition.

A similar pattern as described above was observed for the next group of the samples (Figures 7 and 8). Samples B4 and B6 have uniform microhardness in the axial direction with the average values of 575 and 475 HV, respectively, but sample B3 has a significant difference in microhardness between the upper and lower edges which may be caused by a non-uniform pressing of the powder before electric discharge. Such effects may occur due to the friction of the powder against the walls of the die. The difference in the density distribution for sample B4 is also associated with an unstable consolidation regime. Other samples have the average microhardness values of 475 and 500 HV and density values of

97%. The strengthening effect of nanoparticles is a known phenomenon [21]. However, any significant dependence of the microhardness on the content of yttria had not been identified because of the rather small content of the oxide.

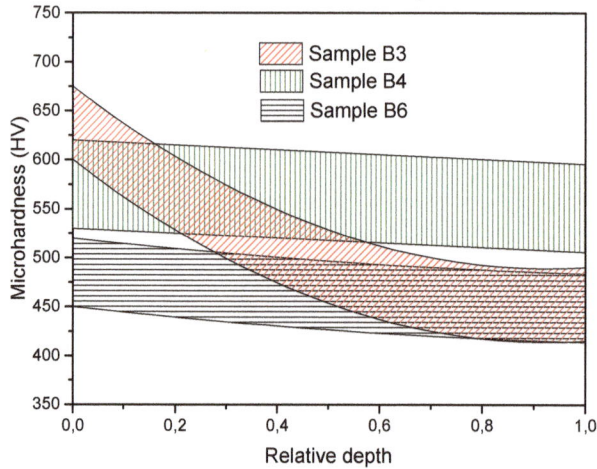

Figure 7. Axial microhardness distribution along samples' depth with 0.3 wt% of yttria addition.

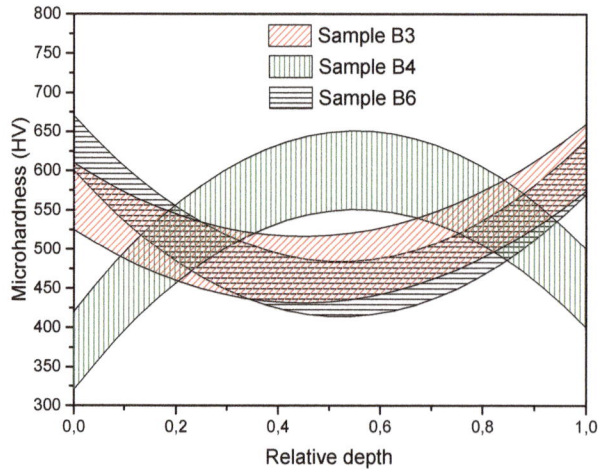

Figure 8. Radial microhardness distribution along samples' depth with 0.3 wt% of yttria addition.

The axial density distribution of the consolidated samples is shown in Figure 9. The experimental data for the two groups of samples containing 0 wt% (samples B5, B7, B8) and 0.7 wt% of yttria (samples B10, B11) were obtained. It is clear that there is no stable density uniformity between layers of the processed samples. Therefore, on average, the difference from layer to layer is from 2% to 3%, which is not a significant enough value to determine any regular patterns of its distribution. There is a slight decrease in density for the layer adjacent to the lower punch for several samples due to the non-uniform pressing conditions before the electric discharge. Despite the fact that the pressing process is double-sided with a loosely attached die, there is some friction of the powder against the wall of the die during pressing. This leads to the non-uniform densification of the powder. Also, no correlation between density and microhardness distribution was observed. This may be due to the fact that different factors independently affect the microhardness and the density of the samples. The factors affecting microhardness may depend not only on the density but also on the phase composition, on the presence of impurities or defects, and on other factors.

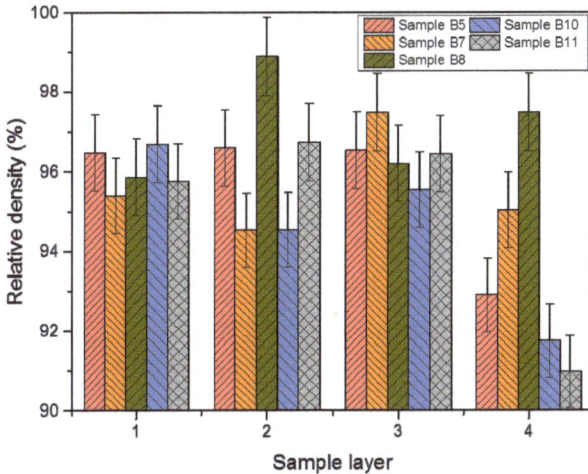

Figure 9. Axial density distribution along depth of the samples.

Microstructure analyses of the consolidated samples were held to determine structure dependence on various factors of the compaction process such as applied pressure and cooling rate after electric discharge. The central area and the periphery of the samples were studied. Figure 10a shows a typical microstructure of the consolidated compacts in the center of the sample. It can be seen that the structure consists of some light color formations which may include agglomerated powder particles with a dark color fine structure between them. The size of the observed agglomerates is 20–50 μm along the minor axis and 100–200 μm along the major axis. The agglomerates are predominantly orientated in the direction perpendicular to the compaction axis of the sample, *i.e.*, they are flattened under pressure during the consolidation process. Figure 10b shows the microstructure at the edge of the sample. It is clear that the porous area reaches the depth of 100–150 μm and then a fairly dense structure is observed.

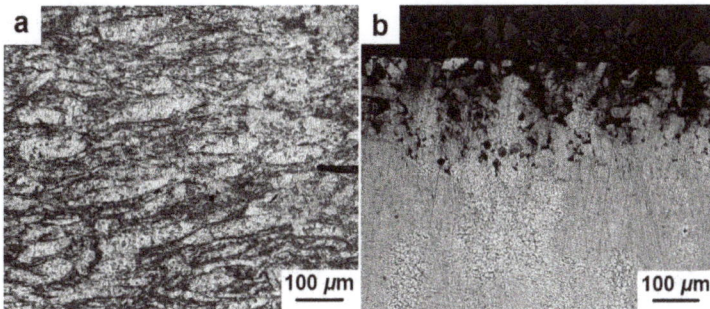

Figure 10. Microstructure of the center (**a**) and edge (**b**) of the consolidated samples.

Figures 11 and 12 indicate microstructure differences between various areas of the samples' volume (center, edge, and side surface). A more detailed examination of the samples' surface allows distinguishing the grain structure and the presence of pores (Figure 11a,b). These pores are substantially elongated and located along grain boundaries. A non-typical structure was also observed primarily at the edge of the samples. Figure 12a shows a porous structure at the edge of the sample adjacent to the punches. Due to the rapid cooling the martensitic structure is formed in these areas. A different structure was observed at the edge of the sample adjacent to the side surface (Figure 12b). It is a mixture of ferrite and perlite with a distance between the perlite plates of around 1–2 μm. This structure could be formed as a result of intensive cooling of the side surface of the sample, but not as rapidly as in the case shown in Figure 12a.

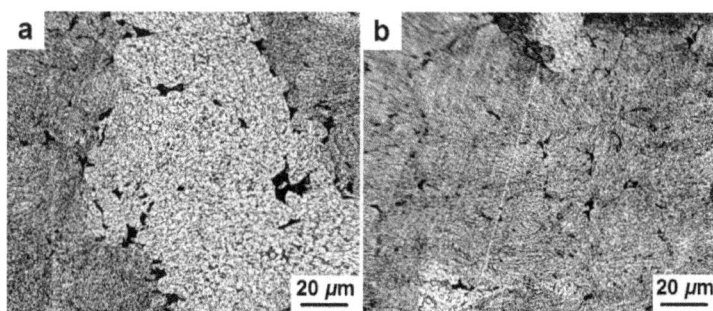

Figure 11. Microstructure of the center area (**a,b**) of the consolidated samples.

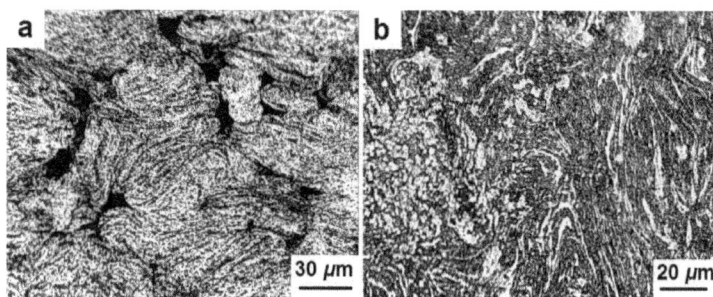

Figure 12. Microstructure of the edge (**a**) and side surface (**b**) of the consolidated samples.

4. Conclusions

Oxide dispersion-strengthened 13Cr-2Mo steel powder was successfully consolidated to near-theoretical density using high voltage electric discharge compaction. Cylindrical samples with relative density from 90% to 97% and dimensions of 10 mm in diameter and 10–15 mm in height were obtained.

It was found that high pressures lead to better interparticle contacts and to the decreasing resistivity of these contacts. This, in turn, leads to the lower energy release and heating of the powder which affects the sintering consolidation mechanism. It is clear that, using high pressures, it is possible to compact samples to high density values only by cold plastic deformation mechanisms; however, the strength and hardness of such samples may be insufficient. Therefore, it is necessary to combine both mentioned mechanisms to obtain an acceptable product. As a result, 200 MPa and 4.2 kV consolidation conditions were chosen as optimal for the compaction of ODS steel powder.

A slight increase of the microhardness at the edges of the samples in the axial and radial directions was observed. During electric discharge, the powder areas located in the vicinity of the punches can densify more intensively due to a higher degree of plastic deformation in comparison with the central parts of the processed specimens. Another reason for this effect may be associated with the high heating and cooling rates of these areas.

No correlation between density and microhardness distribution in the volume of the processed samples was observed. This may be due to the fact that different factors independently affect the microhardness and density of the samples.

It was shown that the edges of the samples adjacent to the punches have a martensitic structure due to rapid cooling in these areas. A different structure was observed at the edges adjacent to the side surface of the processed samples. It is a mixture of ferrite and perlite with a distance between the perlite plates of around 1–2 μm. This structure could be formed as a result of the intensive cooling of the side surface of the samples, but not as rapidly as in the locations of the martensitic phase.

In general, high voltage compaction is an acceptable method for the consolidation of ODS steels. This ultra-rapid process allows obtaining high density compacts, retaining the initial structure with minimal grain growth, and avoiding thermally activated phase transformations. Nevertheless, some heterogeneity of the structure may occur in the boundary layers of the processed samples due to thermal and electromagnetic effects. Therefore, the choice of the optimal parameters of consolidation is required for obtaining samples of acceptable quality.

Acknowledgments: The support of the Department of Science and Education of Russian Federation (Grant Contract 11.G34.31.0051) is gratefully appreciated. The support of RF Ministry of Education and Science, project RFMEFI57514X0003, by the State Program "Science", Project #533 is gratefully acknowledged. The support of the San Diego State University researcher by the US Department of Energy, Materials Sciences Division, under Award No. DE-SC0008581 is gratefully acknowledged.

Author Contributions: I. Bogachev conceived and designed the experiments; performed density, microhardness measurements and microstructure investigations; analyzed the data; wrote the paper; A. Yudin performed the electric discharge consolidation experiments and analyzed the data; E. Grigoryev, I. Chernov, M. Staltsov, O. Khasanov and E. Olevsky supervised the article and participated in the discussion.

Conflicts of Interest: The authors declare no conflict of interest.

References

1. Toualbi, L.; Cayron, C.; Olier, P.; Malaplate, J.; Praud, M.; Mathon, M.-H.; Bossu, D.; Rouesne, E.; Montani, A.; Loge, R.; *et al.* Assessment of a new fabrication route for Fe–9Cr–1W ODS cladding tubes. *J. Nucl. Mater.* **2012**, *428*, 47–53. [CrossRef]

2. Ukai, S.; Fujiwara, M. Perspective of ODS alloys application in nuclear environments. *J. Nucl. Mater.* **2002**, *307*, 307–311, 749–757. [CrossRef]

3. Oksiuta, Z.; Olier, P.; de Carlan, Y.; Baluc, N. Development and characterization of a new ODS ferritic steel for fusion reactor application. *J. Nucl. Mater.* **2009**, *393*, 114–119. [CrossRef]

4. Xu, Y.; Zhow, Z.; Li, M.; He, P. Fabrication and characterization of ODS austenitic steels. *J. Nucl. Mater.* **2011**, *417*, 283–285. [CrossRef]

5. Heintze, C.; Hernández-Mayoral, M.; Ulbricht, A.; Bergner, F.; Shariq, T.; Weissgärber, T.; Frielinghaus, H. Nanoscale characterization of ODS Fe–9%Cr model alloys compacted by spark plasma sintering. *J. Nucl. Mater.* **2012**, *428*, 139–146. [CrossRef]

6. Liu, F.; Liu, Y.; Wen, Y.; Doua, Y.; Zhaoa, D.; Liuc, C.T. Microstructures and mechanical properties of Fe-14Cr-3W-Ti-Y2O3 steel with 1 wt% Cu addition fabricated by a new method. *J. Nucl. Mater.* **2011**, *414*, 422–425. [CrossRef]

7. Sun, Q.X.; Zhang, T.; Wang, X.P.; Fang, Q.F.; Hao, T.; Liu, C.S. Microstructure and mechanical properties of oxide dispersion strengthened ferritic steel prepared by a novel route. *J. Nucl. Mater.* **2012**, *424*, 79–284. [CrossRef]

8. Giuntini, D.; Olevsky, E.A.; Garcia-Cardona, C.; Maximenko, A.L.; Yurlova, M.S.; Haines, C.D.; Martin, D.G.; Kapoor, D. Localized overheating phenomena and optimization of spark-plasma sintering tooling design. *Materials* **2012**, *6*, 2612–2632. [CrossRef]

9.   Oksiuta, Z.; Baluc, N. Effect of mechanical alloying atmosphere on the microstructure and Charpy impact properties of an ODS ferritic steel. *J. Nucl. Mater.* **2009**, *386–388*, 426–429. [CrossRef]

10. Seki, M.; Hirako, K.; Kono, S.; Kihara, Y.; Kaito, T.; Ukai, S. Pressurized resistance welding technology development in 9Cr-ODS martensitic steels. *J. Nucl. Mater.* **2004**, *329–333*, 1534–1538. [CrossRef]

11. Eiselt, Ch. Ch.; Klimenkov, M.; Lindau, R.; Möslang, A. Characteristic results and prospects of the 13Cr–1W–0.3Ti–0.3Y$_2$O$_3$ ODS steel. *J. Nucl. Mater.* **2009**, *386–388*, 525–528. [CrossRef]

12. Zhang, C.H.; Kimura, A.; Kasada, R.; Jang, J.; Kishimoto, H.; Yitao Yang, Y. Characterization of the oxide particles in Al-added high-Cr ODS ferritic steels. *J. Nucl. Mater.* **2011**, *417*, 221–224. [CrossRef]

13. De Bremaecker, A. Past research and fabrication conducted at SCKCEN on ferritic ODS alloys used as cladding for FBR's fuel pins. *J. Nucl. Mater.* **2012**, *428*, 13–30. [CrossRef]

14. Ukai, S.; Kaito, T.; Ohtsuka, S.; Narita, T.; Fujiwara, M.; Kobayashi, T. Production and properties of nano-scale oxide dispersion strengthened (ODS) 9Cr martensitic steel claddings. *ISIJ Int.* **2003**, *43*, 2038–2045. [CrossRef]

15. Yurlova, M.S.; Grigoryev, E.G.; Olevsky, E.A.; Demenyuk, V.D. Electric pulse consolidation of tantalum anodes for electrolytic capacitors. *Inorganic Materials: Applied Research* **2015**, *6*, 267–274. [CrossRef]

16. Yurlova, M.S.; Demenyuk, V.D.; Lebedeva, L. Yu.; Dudina, D.V.; Grigoryev, E.G.; Olevsky, E.A. Electric pulse consolidation: An alternative to spark plasma sintering. *J. Mater. Sci.* **2014**, *49*, 952–985. [CrossRef]

17. Grigoryev, E.G.; Olevsky, E.A. Thermal processes during high-voltage electric discharge consolidation of powder materials. *Scr. Mater.* **2012**, *66*, 662–665.

18. Lee, G.; Yurlova, M.S.; Giuntini, D.; Grigoryev, E.G.; Khasanov, O.L.; Izhvanov, O.; Back, C.; McKittrick, J.; Olevsky, E.A. Densification of zirconium nitride by spark plasma sintering and high voltage electric discharge consolidation: A comparative analysis. *Ceram. Int.* **2015**, *41*, 14973–14987.

19. Bogachev, I.; Olevsky, E.; Grigoryev, E.; Khasanov, O. Fabrication of 13Cr-2Mo ferritic/martensitic oxide-dispersion-strengthened steel components by mechanical alloying and spark-plasma sintering. *JOM* **2014**, *66*, 1020–1026. [CrossRef]

20. Grigoryev, E.G.; Olevsky, E.A.; Yudin, A.V.; Yurlova, M.S. Wave mode high voltage consolidation of powder materials. *Comp. Mater. Sci.* **2014**, *100*, 8–14. [CrossRef]

21. Moghadam, A.D.; Schultz, B.F.; Ferguson, J.B.; Omrani, E.; Rohatgi, P.K.; Gupta, N. Functional metal matrix composites: Self-lubricating, self-healing, and nanocomposites—An outlook. *JOM* **2014**, *66*, 872–881. [CrossRef]

materials

MDPI

Article

Characterizations of Rapid Sintered Nanosilver Joint for Attaching Power Chips

Shuang-Tao Feng [1,2], Yun-Hui Mei [1,2,*], Gang Chen [3], Xin Li [1,2] and Guo-Quan Lu [1,4]

[1] Key Laboratory of Advanced Ceramics and Machining Technology of Ministry of Education, Tianjin University, 135# Yaguan Road, Jinnan District, Tianjin 300350, China; 13207567335@163.com (S.-T.F.); xinli@tju.edu.cn (X.L.); gqlu@vt.edu (G.-Q.L.)
[2] Tianjin Key Laboratory of Advanced Joining Technology and School of Materials Science and Engineering, Tianjin University, Tianjin 300350, China
[3] School of Chemical Engineering and Technology, Tianjin University, Tianjin 300350, China; agang@tju.edu.cn
[4] Department of Materials Science and Engineering, Virginia Tech, Blacksburg, VA 24061, USA
*   Correspondence: yunhui@tju.edu.cn; Tel.: +86-22-2740-8399; Fax: +86-22-2740-5889

Academic Editor: Eugene A. Olevsky
Received: 9 May 2016; Accepted: 27 June 2016; Published: 12 July 2016

Abstract: Sintering of nanosilver paste has been extensively studied as a lead-free die-attach solution for bonding semiconductor power chips, such as the power insulated gated bipolar transistor (IGBT). However, for the traditional method of bonding IGBT chips, an external pressure of a few MPa is reported necessary for the sintering time of ~1 h. In order to shorten the processing duration time, we developed a rapid way to sinter nanosilver paste for bonding IGBT chips in less than 5 min using pulsed current. In this way, we firstly dried as-printed paste at about 100 °C to get rid of many volatile solvents because they may result in defects or voids during the out-gassing from the paste. Then, the pre-dried paste was further heated by pulse current ranging from 1.2 kA to 2.4 kA for several seconds. The whole procedure was less than 3 min and did not require any gas protection. We could obtain robust sintered joint with shear strength of 30–35 MPa for bonding 1200-V, 25-A IGBT and superior thermal properties. Static and dynamic electrical performance of the as-bonded IGBT assemblies was also characterized to verify the feasibility of this rapid sintering method. The results indicate that the electrical performance is comparable or even partially better than that of commercial IGBT modules. The microstructure evolution of the rapid sintered joints was also studied by scanning electron microscopy (SEM). This work may benefit the wide usage of nanosilver paste for rapid bonding IGBT chips in the future.

Keywords: nanosilver; die-attach; current-assisted sintering; rapid joining; characterization

1. Introduction

Insulated gated bipolar transistors (IGBTs) have become important Device for power systems applications such as high-voltage direct current transmission, lamp circuit and variable speed drives, and traction [1,2]. The requirements in size, weight, reliability, durability, ambient temperature, and environment are driving the operation temperatures of power electronics higher than 200 °C [3]. It is known that materials and packaging technologies play more and more important roles in the field of power electronic packaging. In order to reduce the junction temperature of silicon-based IGBT modules, more and more attention has been paid to packaging technologies and materials [4].

In order to avoid the effects of lead, people proposed many lead-free solders, i.e., Ag-Sn or Au-Sn solder [5,6]. Although these modules can work at the high temperature of 200 °C, the short lifetime or high cost still restrict their application. More and more attention has been paid to using nanosilver paste in power electronic industry, especially for high temperature applications [7–10]. Compared with traditional lead-tin and lead-free solder, which are widely used as the die-attach materials for power electronics, low-temperature sintering of nanosilver paste has become a promising lead-free chip joining method. This is attributed to its superior thermal conductivity, electrical conductivity, and reliability because of its high melting point (960 °C) [11,12]. Bai et al. [13] have demonstrated that the fabrication of high temperature devices using nanosilver paste presented superior characteristics over solder joints, including better electrical, thermal and mechanical properties. Ogura et al. [14] found that diode packages made with sintered silver interconnects had electrical and thermal properties equal to those with lead-soldered interconnects.

However, the conventional way to sinter nanosilver requires either a relatively long processing time (up to one hour) under zero pressure or hot pressing, during which the parts are under uniaxial stresses of several megapascals for tens of seconds to a few minutes at the sintering temperature [15–17]. Moreover, this time-consuming process may also cause excessive grain growth [18], which may decrease the mechanical properties of sintered nanosilver and then limit its applications.

A number of novel sintering methods, such as microwave sintering [19], selective laser sintering [20], and electric-current-assisted sintering [21–24], have been put forward to improve the efficiency and properties of sintered materials. Microwave sintering [19] and selective laser sintering [20] are usually used to sinter metals/metal matrix composites and ceramics, while electric-current-assisted sintering can be used for joining materials with high mechanical performance [23]. Recently, a concept of rapid sintering of nanosilver paste using pulsed current for joining, e.g., bus-bar interconnection, has been studied [23]. It has received attention in recent years because of many advantages: extremely high heating rate, 100~1300 °C/s, which could help to bypass the low temperature regime and avoid the aggregation of the nanoparticles; short sintering time, which is a benefit for improving the efficiency; and almost no grain growth, which could lead to the enhancement of mechanical properties of sintered nanosilver. Allen et al. [25] used electrical current assisted sintering (ECAS) to sinter nanosilver on temperature-sensitive photopaper. The conductivity of the sintered nanosilver reached as high as 3.7×10^7 S·m^{-1}. Mei et al. [23] used ECAS to bond copper plates by sintering of nanosilver in less than one second and the sintered joints show high shear strength, i.e., 40 MPa. Extremely high heating rates also make it possible to sinter nanoparticles with insignificant grain growth [26], leading to fine grains, i.e., 300 nm on average, and superior mechanical properties.

Although we had bonded copper plates successfully using the alternative current (AC) by sintering of nanosilver paste before [23] and the properties of the sintered nanosilver by electrical current are good, we only used it for bonding copper plates as bus-bar interconnection. Unfortunately, combination of AC of more than 6.0 kA and pressure of more than 10 MPa had to be used to get a robust sintered nanosilver joint in our previous work. It is doubted whether the method could be used for attaching power chip because the semiconductor chips are not conductive and could not take such high current. The objective is to bond IGBT chips with a substrate by sintering nanosilver paste using electrical current and characterize its mechanical, thermal, microstructural, and electrical properties. In this paper, we have bonded IGBT chips successfully with copper plates electroplated with silver rapidly by sintering of nanosilver paste at low current, i.e., 2.0 kA, and low pressure i.e., 1 MPa. This rapid joining method can be used to form robust die attachment by avoiding excessive current breakthrough power semiconductor chip.

2. Materials and Methods

Figure 1 shows a TEM micrograph of silver nanoparticles in the paste used in this study. The composition of the paste is present in our previous work [23]. These nanoparticles have a wide particle size distribution from 20 to 140 nm. The average size is 50 nm. The nanosilver paste was prepared by mixing selected organics, surfactants, and binders with silver nanoparticles. The organics can prevent aggregation or agglomeration of the silver nanoparticles at low temperatures, e.g., below 200 °C. Once the temperature is increased higher, most of the organics will be burned out. Then, the silver nanoparticles can experience favorable densification by grain boundary diffusion [27]. Evident silver-silver necks could be formed among these particles uniformly.

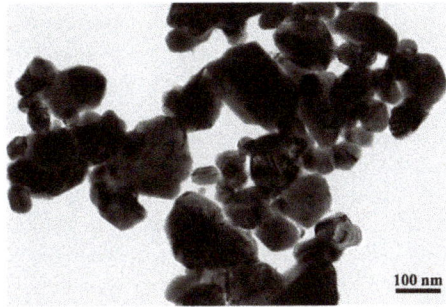

Figure 1. A transmission electron microscopic image of silver nanoparticles.

The substrate used for attaching IGBT chips is made of copper with 5 μm silver coating on the surface. The dimension of the substrate is 23 mm × 15 mm × 1.5 mm. Before printing nanosilver paste, the substrate should be cleaned ultrasonically in alcohol for 30 min. Then a square layer of nanosilver paste (8 mm × 6 mm × 0.09 mm) was stencil-printed on the substrate. An IGBT die (6.5 mm × 4.87 mm × 0.12 mm, 1200 V, 25 A, SIGC32T120R3LE) was picked and placed on the as-printed paste. The specimens were first pre-dried on a heating plate for 0 to 30 min after slowly heating from room temperature to the pre-dried temperature at a heat rate of 5 °C/min in order to remove most of the solvents in the paste. Then the pre-dried specimen was sintered using pulsed current for several seconds, i.e., 90 s, 120 s, 150 s, and 180 s, under a low pressure. The schematic diagram of the current sintering process is shown in Figure 2. The power source is able to provide both pulsed and continuous electric current up to 10 kA. A self-design fixture was used to position the pre-dried specimen. A piece of SiC was used here as an insulation to avoid current flowing elsewhere and reduce the heat dissipation to the base. We used SiC in this work because the SiC is rigid and insulative enough with relatively low thermal conductivity. The temperature distribution of the as-printed nanosilver paste during sintering was measured by an infrared radiation camera, which was placed in front of the sintering equipment. A clear image of the temperature distribution contour of the nanosilver joints could be achieved by adjusting the focal length and the height of the tripod. A typical as-sintered specimen is also shown in Figure 2. The specimen was sintered in air under the combined condition of the sintered current of 2.0 kA, the current-on time of 150 s, and the assisted pressure was 1 MPa.

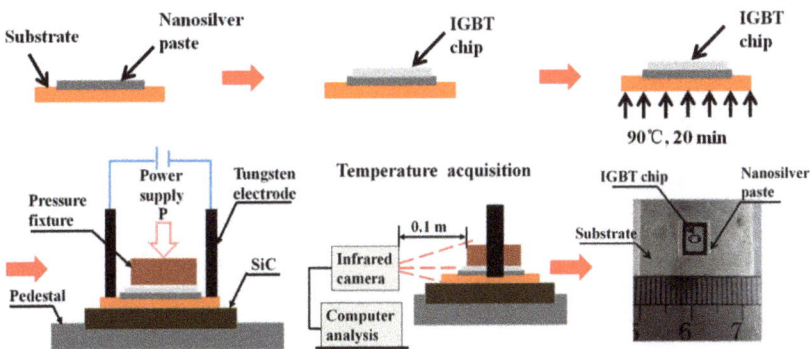

Figure 2. Schematic diagram of pulse-current-assisted sintering process.

It is critical to monitor the temperature profile of the sintering process in order to get robust as-sintered nanosilver joints. However, it is inconvenient to measure the temperature profile by conventional ways, e.g., thermal couple, because the thermal couple should be mounted on the surface of region-of-interests (ROI) and could only measure the temperature locally. An infrared radiation (IR) camera (Guangzhou SAT Infrared Technology Co., Ltd., SAT G90, Guangzhou, China) was adopted to measure the temperature distribution and evolution of the rapid current-assisted sintering process. The accuracy of the results by the IR camera should be highly dependent on the determination of emissivity. As a result, the reference channel method [28] was used to obtain accurately the emissivity of the average temperature of the nanosilver layer and surrounding area (ROI). A specific thermal stable black paint (Botny, 550 °C High temperature paint, Guangzhou, China) with the constant emissivity, i.e., 0.95, was used to cover the ROI of the pre-dried specimen. It is difficult to measure the temperature profile of the nanosilver layer, because evaporation of the solvent along with burnout of most of the organics in the paste, and densification of the silver particles during the sintering process will cause shrinking of the thickness of the nanosilver layer [29].

Figure 3 shows the cross section of the sintered joint without introducing fixing pressure. There is a significant gap between the IGBT chip and the sintered nanosilver. It is likely that the rapid temperature ramping by the electrical current heating caused the abrupt outgassing of the organics in the paste. Significant force was induced by the abrupt outgassing of the organics and could drive the IGBT chip away from the paste. As a result, the gap could be generated at the interface between the IGBT chip and the paste in this case. The gap should hinder the further atomic migration of silver during sintering to generate robust joint. In order to get rid of the unexpected gap, a fixture was designed to provide ~1 MPa pressure on the chip in this work.

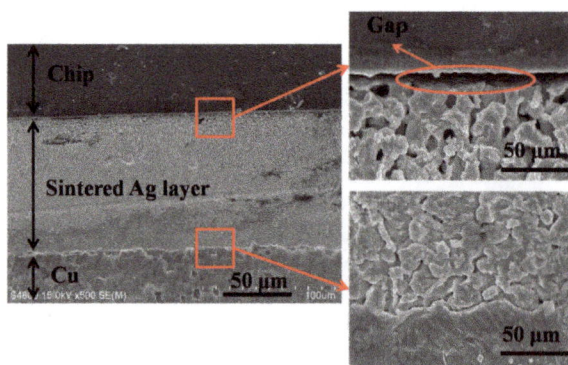

Figure 3. Scanning electron microscopic images of cross section of sintered silver joint without pressure.

The electrical properties of the as-sintered IGBT assemblies, i.e., switching on/off performance, were characterized by double pulse testing [30] in order to verify the feasibility of the rapid sintering method for bonding power chips. Figure 4a shows the circuit of the double pulse testing schematically. Figure 4b show the electrical connects of the double pulse testing.

Gain/particle size of the sintered nanosilver was also measured to correlate with the sintering process. The most widely used method of average grain/particle size measurement is the mean lineal intercept. In order to prepare samples for the measurement, the fracture surface of sintered nanosilver was dipped for 4 s in an etching solution of 30 vol % ammonia (NH_4OH), 43 vol % hydrogen peroxide (H_2O_2), and 27 vol % distilled water (H_2O), and then washed by distilled water. The microstructures of etched fracture were observed by SEM to reveal the particle size. The etched fracture surface was analyzed at different regions at least three times by SEM. According to ASTM E112-96, the mean lineal intercept length is the average length of a line segment that crosses a sufficiently large number of grains. It is proportional to the equivalent diameter of a spherical grain. The mean lineal intercept length is determined by laying a number of randomly placed test lines on the image and counting the number of times that grain boundaries are intercepted. Mathematically, it is defined as:

$$\overline{L_L} = \frac{1}{N_L} = \frac{L_T}{PM} \tag{1}$$

where N_L is the number of intercepts per total length of the test lines L_T; P is the total number of grain boundary intersections and M is the magnification. All grains/particles in each SEM pictures were measured with a plurality of lines to obtain the average size in this work.

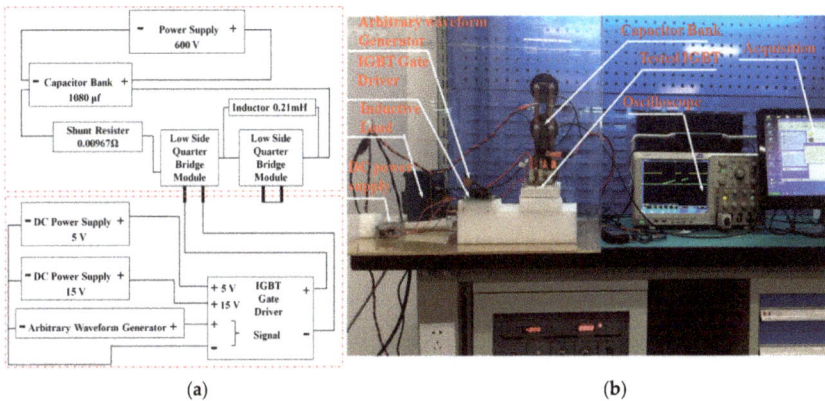

(a) (b)

Figure 4. (a) Schematic diagram of a circuit; (b) a picture of electrical connects of the double pulse testing.

A summary of sintering conditions using electrical current is listed in Table 1. At least three samples were prepared for each condition. The effects of electrical current, current-on time, pre-drying temperature, and pre-drying time on robustness of the sintered IGBT assemblies were evaluated in this work.

Table 1. Summary of sintering conditions using pulsed electrical current.

Condition No.	Current (kA)	Current-on Time (s)	Pre-Drying Temperature (°C)	Pre-Drying Time (min)
1	1.2	150	90	20
2	1.6	150	90	20
3	2.0	150	90	20
4	2.4	150	90	20
5	2.0	90	90	20
6	2.0	120	90	20
7	2.0	180	90	20
8	2.0	150	60	20
9	2.0	150	120	20
10	2.0	150	150	20
11	2.0	150	90	0
12	2.0	150	90	10
13	2.0	150	90	30

Thermal properties of these sintered samples were measured by thermal gravimetric (TG) in air at different final temperature and time with the heating rate of 5 °C/min.

A die-shear tester (XTZTEC Condor 150) was used to measure the shear strength of the as-sintered specimens at a displacement rate of 4×10^{-4} m/s. The thermal resistance of the IGBT assembly using nanosilver paste was characterized by a self-developed thermal impedance measurement system [31]. The fracture surface and the cross-section of the joint were analyzed by scanning electron microscopy (SEM).

The void ratio of the joint was measured by X-ray computed micro-tomography (μ-CT). To identify the void regions from the CT images, an appropriate threshold should be determined. Regions with pixels below this threshold are treated as voids. The details of the methods can be found in the references [32–34]. Since the minimum void size that can be detected by the X-ray tomography is 10 μm, we defined a void in this work as defects that are larger than 10 μm. Comparison of the microstructures and the voids of all the sintered joints were discussed to clarify the relationship among processing, performance, and microstructures.

3. Results

3.1. Temperature Profile

Figure 5a shows the temperature variation of joint during the sintering process under the current of 2.0 kA with different current-on time, i.e., 90 s, 120 s, 150 s, and 180 s. The temperature of the joint rises rapidly at the beginning of the process, i.e., 25 s, with a heating rate of >20 °C/s because of massive instant Joule heat. The heating rate is constant before the peak temperature reaches ~450 °C regardless of the current-on time. Figure 5b shows that the peak temperature could reach almost 550 °C once the sintering current is 2.4 kA. It is concluded that the heating rate and the peak temperature are only dependent on the sintering current and independent of the current-on time.

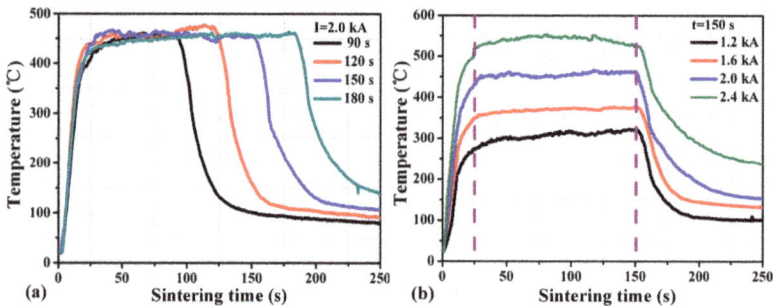

Figure 5. Temperature profile of nanosilver joints sintered at different: (a) current-on time; (b) sintering current.

3.2. Die-Shear Strength and Thermal Resistance

Die-shear strength is one of key factors affecting the mechanical performance and reliability of die attachment [35]. We had realized robust sintered joint for bonding copper plates, i.e., 25 mm^2, by current-assisted sintering of nanosilver paste [7]. It is essential to achieve die-shear strength as high as the conventional solders, hot-pressing sintered nanosilver, i.e., 30 MPa. Furthermore, the thermal property of the IGBT assembly is also important to guarantee the production consistency and reliability, especially for high temperature and high power applications. An improved transient thermal impedance (Z_{th}) measurement system was self-developed [31], using the electrical method with V_{ge} of the IGBT as a temperature-sensitive parameter.

The heat of the sintering process is joule heat which is generated when current flows through the substrate, and it can be expressed as $Q = I^2Rt$. The sintering current and current-on time have great influence on the temperature. Figure 6a shows that the average thermal resistance and shear strength of the IGBT assemblies is strongly dependent on the sintering current and current-on time.

The average thermal resistance of the sintered samples decreases as the current-on time increases. At the same time, the average thermal resistance decreases with increasing the sintering current from

1.2 kA to 2.0 kA. When increasing the sintering current from 2.0 kA to 2.4 kA, however, the average thermal resistance increases slightly. The shear strength of the IGBT assemblies increases as the sintering current and the current-on time increases. The die-shear strength is only ~12 MPa once the current is 1.2 kA and the current-on time is 150 s. In this case, the thermal resistance of the IGBT assembly could be as large as ~0.5 °C/W. Furthermore, the die-shear strength is less than 10 MPa, even when the current increases to 2.0 kA with the current-on time of 90 s. It is likely that the organics could not be burnt out adequately and densification of the silver particles is insufficient once the sintering current or the current-on time is at low levels [23]. If a large amount of organics remained in the paste, strong bonds could not form because the residual organics may hinder atomic inter-diffusion of the silver particles and heat conduction [36,37]. Consequently, the larger sintering current and the longer current-on time could accelerate volatilization, decomposition, or ablation of the organics [7]. Inter-granular diffusion happens between silver particles at high temperatures and the higher current magnitude accelerates the diffusion of silver atoms as well as density. A large amount of heat and local high temperature increase the rate of diffusion of silver atoms and then benefit forming a clear neck between particles, which is consistent with Akada et al. [36].

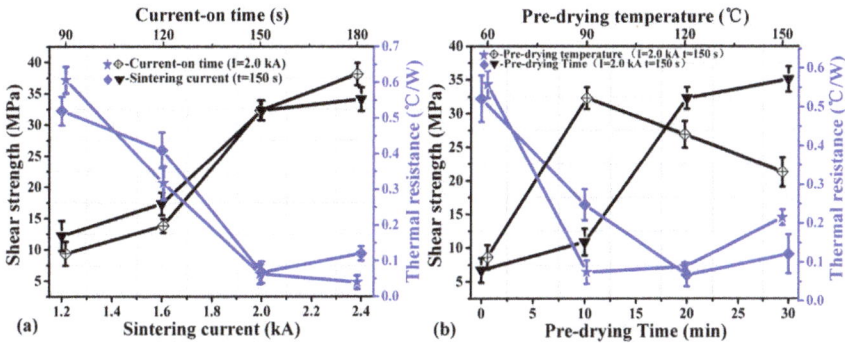

Figure 6. Shear strength and thermal resistance of nanosilver joints sintered: (**a**) at different current (current-on time of 150 s, pre-drying temperature of 90 °C, and pre-drying time of 20 min), at different current-on time (sintering current of 2.0 kA, pre-drying temperature of 90 °C, and pre-drying time of 20 min); (**b**) at different pre-drying temperature (current-on time of 150 s, sintering current of 2.0 kA, and pre-drying time of 20 min), at different pre-drying time (current-on time of 150 s, sintering current of 2.0 kA, and pre-drying temperature of 90 °C).

The larger sintering current also resulted in the larger heating rate, which should be beneficial to reduce non-densification diffusions, i.e., surface diffusion, by bypassing the low-temperature regime instantaneously [36]. The non-densification diffusion is able to consume the driving force for further densification diffusion because the driving force for sintering of nanosilver paste is the tendency to reduce the free energy of the Ag nanoparticles, accomplished by material transport from high energy site to lower one.

Moreover, the extremely large heating rate is prone to form a large amount of twins, which increases the thermal conductivity of the sintered nanosilver [38]. Therefore, the die-shear strength of the IGBT assembly could reach as high as ~35 MPa and the thermal resistance could reduce to only 0.06 °C/W in the condition of 2.0 kA and 150 s.

Figure 6b shows that the effect of pre-drying temperature and pre-drying time on the average die-shear strength and the thermal resistance of the sintered silver joints. It can be seen that the die-shear strength increases and the thermal resistance decreases with increasing pre-drying temperature from 60 °C to 90 °C. However, when increasing the pre-drying temperature from 90 °C to 150 °C, the die-shear strength decreases and the thermal resistance increases. The thermal resistance was 0.52 °C/W, 0.25 °C/W, 0.07 °C/W, and 0.12 °C/W under the pre-drying time of 0 min, 10 min, 20 min, and 30 min, respectively. The die-shear strength increases with increasing pre-drying time. It is likely that a large amount of organics was burned out once the temperature reached 90 °C or even higher because a significant weight loss at ~90 °C, as shown in Figure 7b. Therefore, the pre-dried nanosilver paste became too dry to be deformed and wet due to the burning out of the large amount of the organics when the pre-drying temperature increased from 90 °C to 150 °C. Many defects or air gaps might be present in the interface between the over-dried nanosilver paste and the power chip and hinder further atomic diffusion of silver particles that is the bonding mechanism of sintering of the nanosilver paste although the fixing pressure is helpful to reduce the air gaps at the interface to some extents. The die-shear strength and the thermal properties reduced consequently. The fixing pressure of 1 MPa is too small to avoid most of the defects and the air gaps if the pre-dried paste is too dry and not deformable. It is also not recommended to increase the fixing pressure to a much higher value, e.g., 10 MPa, because there is a risk damaging the chips under such high pressure. It is concluded that the pre-dried temperature is critical to the bonding quality of the sintered nanosilver paste using electrical current and should not be higher than ~90 °C based on the TG results.

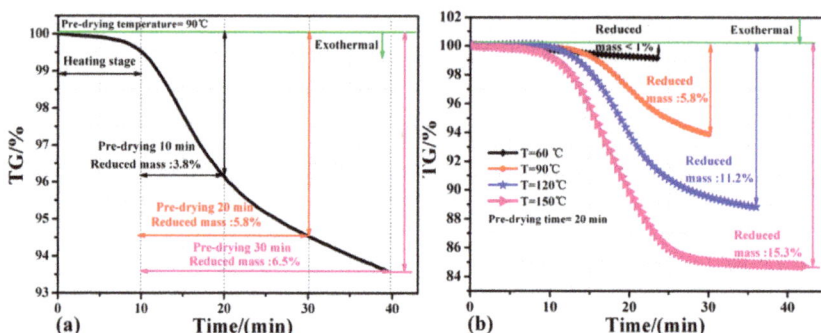

Figure 7. Thermal gravimetric trace of nanosilver paste: (**a**) heated to 90 °C for different durations; (**b**) heated to different temperatures for 20 min at a rate of 5 °C/min.

In order to determine the appropriate pre-drying conditions, thermal properties of these sintered samples were measured by thermal gravimetric (TG) in ambient atmosphere with different pre-drying temperature and different pre-drying time at the heating rate of 5 °C/min. The results are shown in Figure 7. Figure 7a shows that with the increase of pre-drying time the more mass loses of the paste. The mass reduction of the paste with a pre-drying time of 30 min is only 0.7 wt % more than that of the paste with a pre-drying time of 20 min at 90 °C. It indicates that there is no need to prolong the pre-drying time to more than 20 min. The conclusion is also supported by the results of die-shear strength and thermal resistance, as shown in Figure 6b.

Figure 7b shows that the mass reduction of the paste is less than 1%, ~5.8%, ~11.2%, and ~15.3% when pre-drying at 60 °C, 90 °C, 120 °C, and 150 °C for 20 min, respectively. It should be noted that it takes several minutes to heat the paste to the pre-dried temperatures before the temperature remains constant for 20 min. For example, in the case of 90 °C for 20 min as shown in the red line of Figure 7b, it takes 13 min to heat the paste from ambient temperature to 90 °C. As a result, the process lasts 33 min in total. It is also worth noting that the mass reduction of the paste increases to 11.2 wt % and 15.3 wt % once the pre-dried temperature increases to 120 °C and 150 °C, respectively. Such great reduction was due to the evaporation of more organics in the paste and likely led to the defects or delamination at the interfaces among the IGBT, the paste, and the substrate. It is suggested that the evaporation of the organics should be controlled as ~6 wt % in order to improve the joint quality. The conclusion is also supported by the results of die-shear strength and thermal resistance, as shown in Figure 6b.

3.3. Electrical Properties

In order to verify feasibility of this pulse-current-assisted-sintering method for bonding IGBTs, it is essential to study static and dynamic characteristics of the IGBT assembly by the pulse-current-assisted-sintering method. The experimental study herein is based on Infineon IGBT chips SIGC32T120R3LE. The chip parameters can be found in details in the datasheet [39]. The switching on and off behavior was characterized by a double-pulse testing method [40]. Both the turn-on and the turn-off gate resistance are 20 Ω. The inductive load during the measurement is set as 200 nH. The collector voltage, gate voltage, and collector current flowing through the IGBTs were measured by an oscilloscope with high-voltage probes. The measured static and dynamic results are compared with the ones of the datasheet, as listed in Table 2. The switching on and off behavior including the collector voltage V_{ce}, the collector current I_{ce}, and the gate voltage V_{ge} of the IGBT assembly by the rapid sintering of nanosilver paste is presented in Figure 8 (turn-on) and Figure 9 (turn-off).

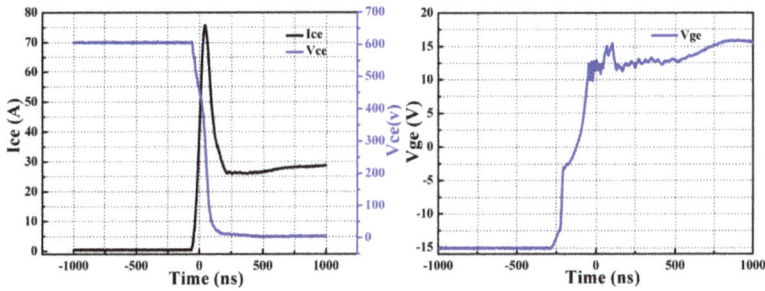

Figure 8. Turn-on behavior of an IGBT at V_{DC} = 600 V, I_C = 25 A, and T = 25 °C. (V_{ce} 100 V/div, I_{ce} 20 A/div, V_{ge} 5 V/div, and time 0.25 μs/div.)

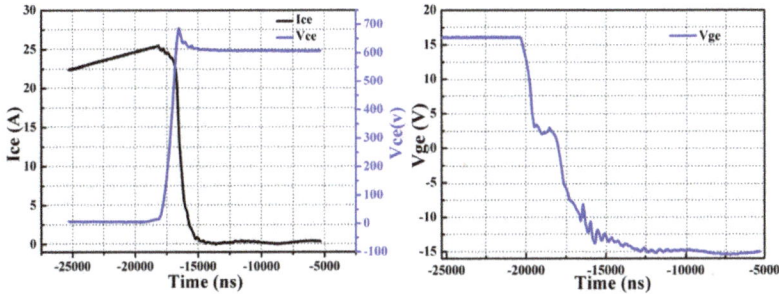

Figure 9. Turn-off behavior of an IGBT at V_{DC} = 600 kV, I_c = 25 A, and T = 25 °C. (V_{ce} 100 V/div, I_{ce} 20 A/div, V_{ge} 5 V/div, and time 0.25 μs/div.)

Table 2. Static and dynamic performance comparison between measured results and values of datasheet.

IGBT (SIGC32T120R3LE) 1200 V/25 A	I_{ces} (uA)	$V_{ce(sat)}$ (V)	$t_{d(on)}$ (ns)	t_r (ns)	E_{on} (mJ)	$t_{d(off)}$ (ns)	t_f (ns)	E_{off} (mJ)
Measured results	0.02	1.73	62	14	1.11	336	114	1.44
Datasheet	Max. 3.48	Typ. 1.70	Typ. 90	Typ. 30	Typ. 2.4	Typ. 420	Typ. 70	Typ. 1.8

It is evident that both the measured I_{ces} and $V_{ce(sat)}$ are comparable or even better than the typical values of commercial Device, indicating that the IGBT assembly by the pulse-current-assisted-sintering method is practicable. The IGBTs were not damaged due to such high heating current as others may concern.

The switching on time is defined as the sum of turn-on delay time, $t_{d(on)}$, and rising time, t_r. The switching off time is defined as the sum of turn-off delay time, $t_{d(off)}$, and falling time, t_f. The switching on and off time are 76 ns and 450 ns, respectively. This is consistent with the fact that the switching off time is usually at level of several nanoseconds and an order of magnitude higher than the switching on time [41].

Moreover, the total IGBT loss is defined as the sum of IGBT turn-on and turn-off losses. For the switching loss evaluation of the IGBT assembly, a total collector current of 25 A is applied. It is worth noting that the overshoot of the collector voltage is less than 50 V during switching on. However, there is a significant overshoot in the collector current. The maximum collector current during switching off could reach up to ~75 A, which is two times higher than the rated collector current. It was likely that the DBC substrate of the IGBT assembly had been oxidized locally, especially the positions close to the heating electrodes, due to such high heat current. The generated copper oxide should increase the resistance because the electrical resistivity of the copper oxide is much larger than those of both copper and silver. The skin effect of the copper metallization, which could be expressed as $\delta = [\rho / (\pi \times \mu \times f)]^{1/2}$, becomes significant at high frequency. In the above equation, δ is skin depth; ρ is electrical resistivity; μ is permeability; and f is operating frequency. As a result, the skin depth should increase as well as the parasitic inductance once the electrical resistivity increases. Then the variation in parasitic inductance led to the overshoot in the collector current of the IGBT assembly during switching off. The turn-on and turn-off loss are 1.11 mJ and 1.44 mJ, respectively. Unfortunately, there is no typical value of the switching characteristics in the datasheet because switching characteristics is depending strongly on module design and mounting technology and can therefore not be specified for a bare die. Thus, we compared our results with the typical values of a commercial IGBT module (MMG25H120XB6TN, MacMic Co., Ltd., Changzhou, China) [42].

4. Discussion

4.1. Fracture Surface

The fracture microstructures of the sintered nanosilver joint at different sintering conditions were investigated by SEM, as shown in Figure 10. Figure 10a shows that the higher the current, the more significant the elongated shape on the fracture surface of the specimen. When the current is less than 2.0 kA, there is no elongated dimple that can be interpreted. It was reported previously that obvious plastic deformation should appear on fracture surface of the sintered nanosilver joint with relatively high shear strength [43]. The fracture failure is a kind of cohesive failure rather than adhesive failure. Then, the shear strength of the sintered nanosilver joint is close to that of soldering joints, e.g., PbSn, AuGe12, and ZnAl5, and pressure assisted sintered joint, i.e., >30 MPa [43,44]. The elongated dimples mean the larger fracture deformation or fracture strain of the sintered nanosilver. If significant elongated dimples are present, the shear strength of the sintered nanosilver could reach at least 30 MPa [23]. The fracture surfaces of the IGBT assemblies that were sintered with different current-on time under the same current of 2.0 kA are shown in Figure 10b. Almost no elongated dimple and plastic flow can be observed in the cases of 90 s and 120 s. Therefore, we considered low shear strength could be used to explain "no elongated dimple". However, significant plastic deformation can be found in the case of 150 s and 180 s. The shear strength of sintered joints with the current-on time of 150 s and 180 s could, therefore, reach 33.2 MPa and 38.1 MPa, respectively.

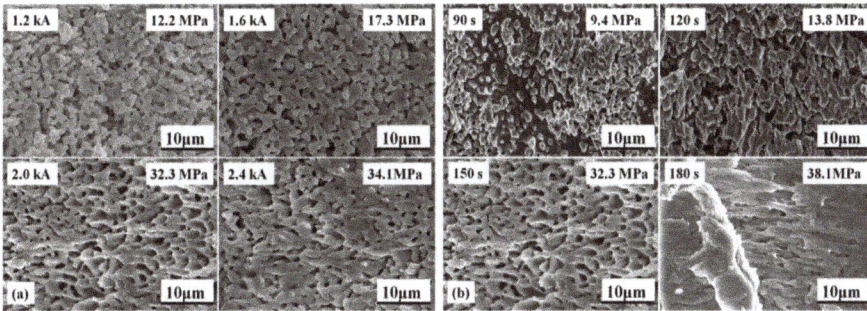

Figure 10. Comparison of microstructures of fracture surface of specimens sintered: (**a**) at different currents (current-on time of 150 s); (**b**) at different current-on times (sintering current of 2.0 kA).

4.2. Void Distribution

Void ratio is also a critical factor that affects die-shear strength and thermal properties of the joints [32]. X-ray micro-tomography (μ-CT), which could identify the void distribution nondestructively based on the criterion of contrast gradient [39], was used to explore the voids in the sintered nanosilver joints of the IGBT assemblies with different sintering current. The right regions as shown in the Figure 11 represent voids or defects in the sintered joints. The area ratio of the right regions could be calculated as the void ratio [33]. More heat could be accumulated at the voids area and then caused the higher junction temperature [34]. During the sintering process initial voids can be transformed into the void with the volatilization of organic matter. Figure 11 shows the microstructures of the sintered nanosilver joint using different sintering current by μ-CT. It can be seen that with an increase of the current, the voids of the joint decreased. Probably because with the temperature increase organic matter has been completely volatile, grain boundary diffusion and lattice diffusion occurred under high temperature to achieve rapid densification of solder paste and the densification process leads to a decrease in the number of voids. Therefore, the density of the sintered joints was enhanced by increasing the current for sintering. It is likely that voids, which could impact the thermal properties of the die attachment greatly [45], are easily formed in the sintered joint and at the interface between the IGBT and the die attachment during the rapid sintering process [36].

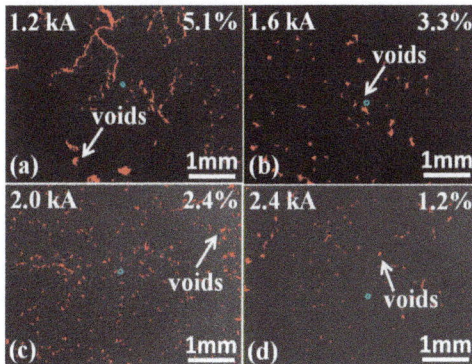

Figure 11. X-ray computed micro-tomography of specimens sintered at: (**a**) 1.2 kA; (**b**) 1.6 kA; (**c**) 2.0 kA; (**d**) 2.4 kA.

4.3. Cross Sections

Figure 12 shows SEM images of cross sections of the sintered silver joints. The as-sintered bondline thickness of the joint increased when prolonging the pre-dried time at the same sintering current of 2.0 kA. It is proven that the as-dried paste at 90 °C for both 0 min and 10 min is too soft to take the fixing pressure, so that the paste could be squeezed out greatly and the as-sintered bondline thickness reduced to 13 μm and 21 μm non-uniformly, respectively. It is reasonable that the die-shear strength of the joints with the pre-dried time of 0 min and 10 min is less than 10 MPa. Once the pre-drying time was prolonged to more than 20 min, the as-sintered joints have dense and uniform bondlines and the as-dried paste could be squeezed out. The as-printed paste is 90 μm thick compared with the as-sintered bondline of less than 30 μm. The bondline shrinkage was due to the densification of silver nanoparticles. However, pressure-assisted sintered nanosilver is usually shrunk to half of the as-printed bondline thickness [46]. The great shrinkage of the sintering paste using electrical current indicates much higher driving force for the densification of the silver nanoparticles compared with that of conventional hot-pressing sintering of nanosilver paste since the processing time is much shorter this way.

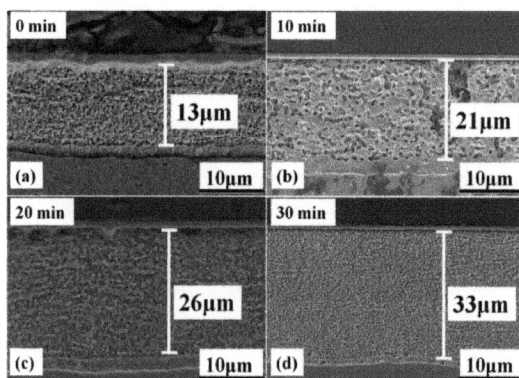

Figure 12. As-sintered bondline thickness of sintered nanosilver with different pre-dried time of: (**a**) 0 min; (**b**) 10 min; (**c**) 20 min; (**d**) 30 min.

The thermal performance and reliability of die attachment greatly depend on the joint density. Enough current-on time is critical to the formation of good bonds at the interfaces between the sintered silver joint and die/substrate [41]. Furthermore, based on Ivensen's sintering theory, the densification in the sintering process could be considered as elimination of crystal defects [47]. The longer dwelling time on the peak temperature, which was proven to be only dependent on the sintering current herein, can promote evaporation of organics in the paste, and necking/nucleation of the silver nanoparticles; thereby, the relative density of the sintered silver joints can be enhanced [48]. The conclusion is consistent with the variation of the average thermal resistance of the IGBT assemblies, which decreases as the current-on time increases from 90 s to 180 s. Therefore, it should be preferable to increase the current-on time to maintain the sintering temperature in order to reduce the crystal defects and accelerate the densification of the die-attach layer.

Moreover, the pore size distribution in the sintered silver joints is shown in Figure 13. Figure 13c,d shows that the pores are small and round in shape, which may be beneficial to the thermal resistance of the die attachment [45]. However, larger and more irregular pores emerged in the sintered silver joints, as can be observed in Figure 13a,b. The shorter surface diffusion could hinder further densification and impact the thermal resistance of die attachment ultimately. It is worth noting that the larger and more irregular pores could impact the long-term reliability of the sintered joints because cracks are easily formed in the vicinity of these pores by stress concentration [49]. The current-on time of 150 s is recommended considering both the thermal performance and reliability of the die attachment.

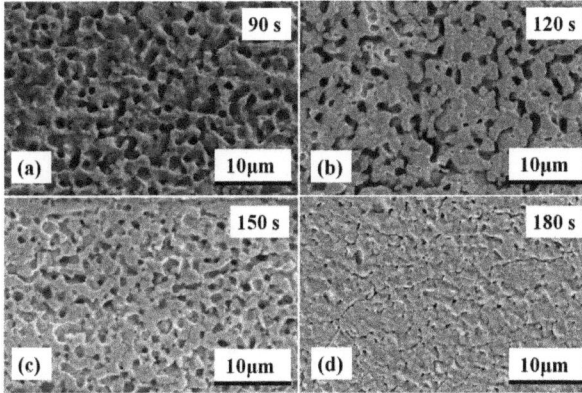

Figure 13. Pore distribution of sintered silver joints sintered at different current-on time of: (**a**) 90 s; (**b**) 120 s; (**c**) 150 s; (**d**) 180 s.

4.4. Evolution of Particle/Grain Size

In order to verify that short sintering time could refine grain, we have to deduce the relationship between grain growth and sintering time. The grain boundary migration rate can be expressed as the following equation [50]:

$$V_b = M_b \times F_b \tag{2}$$

where M_b stands for the mobility of grain boundary; and F_b is the driving force. The M_b is temperature dependent as the following:

$$M_b = (D_b \times \Omega)/(K \times T \times W_b) \tag{3}$$

where D_b is the diffusivity of grain boundary; W_b is boundary width; Ω is atomic volume; K is Boltzmann constant; and T is absolute temperature. The driving force, F_b, can be expressed as the following:

$$F_b = A \times \sigma_b/D_g \tag{4}$$

where σ_b is surface energy; and D_g is average particle size. Grain boundary diffusion could lead to grain growth, as mention above. Therefore, the grain boundary migration rate is proportional to grain growth rate, i.e., dD_g/dt.

$$dD_g/dt = A \times \sigma_b \times D_b \times \Omega/(K \times T \times W_b \times D_g) \tag{5}$$

We integrate Equation (4) and attain the following:

$$D_g{}^2 - D_{g0}{}^2 = k \times t \tag{6}$$

where D_{g0} is the grain size when t equals zero. These equations are only deduced for analyzing the relationship between grain size and sintering time in a qualitative way, so the constants/parameters in these equations are not determined. As a result, we could conclude that the grain growth should dominate the densification when prolonging the current-on time, especially for 180 s when significant grain coarsening could be observed in Figure 14. Figure 15 shows that the grain size on average increases from ~100 nm to more than 500 nm. The apparent strength and the grain size follow a Hall–Petch [51] type behavior:

$$\sigma = \sigma_0 + kd^{-a} \tag{7}$$

where σ is the flow strength; σ_0 and k are size-independent constants; and a is an exponent typically between 0.5 and 1. In the microscopic view, the increase of grain size could lead to the dissolving of the grain boundary. However, high values for yield stress were considered to be related to the effect of increased grain boundaries, providing additional obstacles for movement of lattice dislocations [51]. Therefore, grain refinement could lead to strength enhancement.

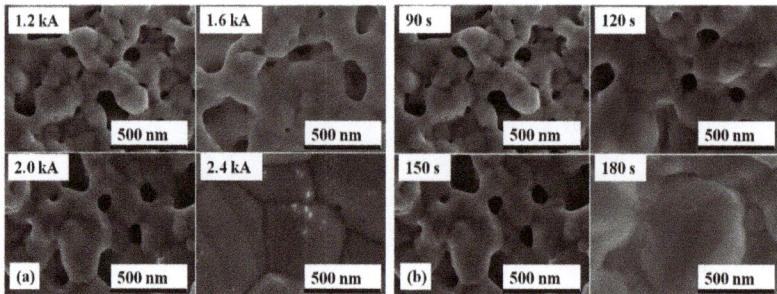

Figure 14. Particle/grain size of nanosilver joints sintered: (**a**) at different currents (current-on time of 150 s); (**b**) different current-on time (sintering current of 2.0 kA).

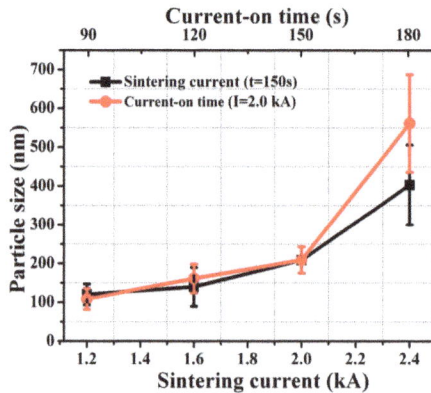

Figure 15. Particle/grain size distribution of nanosilver joints sintered at different currents (t = 150 s) and different current-on times (I = 2.0 kA).

Furthermore, grain refinement of nanocrystals leads to an increase in resistance to failure under stress-controlled fatigue, whereas a deleterious effect was found on the resistance to fatigue crack growth [52]. It is, therefore, not recommended to prolong the current-on time to 180 s. The die-shear strength of the sintered joint under the current of 2.0 kA for 150 s could be as high as 32.3 MPa. One hundred fifty seconds is acceptable as the optimized heating time for the current-assisted sintering of nanosilver paste under the current of 2.0 kA.

5. Conclusions

Compared with the traditional hot-pressing way to sinter nanosilver paste, which takes than half an hour, we are able to sinter the paste to bond IGBT chips in a much shorter time, i.e., ~150 s

Materials **2016**, *9*, 564

with the help of a pulsed current of ~2.0 kA. We obtained robust sintered joints as die attachment for bonding (1200 V, 25 A) IGBT chips (Infineon, SIGC32T120R3LE) using this rapid method. In this way, the die-shear strength and the thermal resistance of the die attachment could reach up to 30–35 MPa and 0.07 °C/W, respectively. Both the static and dynamic electrical performances are comparable or even partially better than that of commercial IGBT modules, indicating that the IGBT assembled using the pulse-current-assisted-sintering method is practical and will not damage power chips due to such high heating current, as other methods might. This work may guide a rapid process to use nanosilver paste for bonding IGBT chips in the future.

Acknowledgments: The authors would like to gratefully acknowledge the financial support from the National Natural Science Foundation of China (under Grant 2015AA034501), the Tianjin Municipal Natural Science Foundation (under grant 13ZCZDGX01106), and the China Postdoctoral Science Foundation (under Grants 2014M551021 and 2015T80219).

Author Contributions: Shuang-Tao Feng and Yun-Hui Mei conceived, designed, and done the experiments; Shuang-Tao Feng, Yun-Hui Mei, Gang Chen analyzed the data; Shuang-Tao Feng, Yun-Hui Mei, Gang Chen, Xin Li, and Guo-Quan Lu wrote the paper.

Conflicts of Interest: The authors declare no conflict of interest.

References

1. Li, K.; Tian, G.Y. State detection of bond wires in IGBT modules using eddy current pulsed thermography. *IEEE Trans. Power Electron.* **2014**, *29*, 5000–5009. [CrossRef]
2. Choi, U.M.; Blaabjerg, F.; Lee, K.B. Study and handling methods of power IGBT module failures in power electronic converter systems. *IEEE Trans. Power Electron.* **2015**, *30*, 2517–2533. [CrossRef]
3. Göbl, C.; Faltenbacher, J. Low temperature sinter technology die attachment for power electronic applications. In Proceedings of the 6th International Conference on Integrated Power Electronics Systems (CIPS), Niagara Falls, ON, Canada, 11–13 October 2010; pp. 1–5.
4. Oh, H.; Han, B.; McCluskey, P. Physics-of-failure, condition monitoring, and prognostics of insulated gate bipolar transistor modules: A review. *IEEE Trans. Power Electron.* **2015**, *30*, 2413–2426. [CrossRef]
5. Riedel, G.J.; Schmidt, R.; Liu, C. Reliability of large area solder joints within IGBT modules: Numerical modeling and experimental results. In Proceedings of the 7th International Conference on Integrated Power Electronics Systems (CIPS), Boston, MA, USA, 5–8 March 2012; pp. 1–6.
6. Tanimoto, S.; Matsui, K. High Junction Temperature and Low Parasitic Inductance Power Module Technology for Compact Power Conversion Systems. *IEEE Trans. Electron Device* **2015**, *62*, 258–269. [CrossRef]
7. Mei, Y.H.; Cao, Y.J.; Chen, G. Rapid sintering nanosilver joint by pulse current for power electronics packaging. *IEEE Trans. Device Mater. Reliab.* **2013**, *13*, 258–265. [CrossRef]
8. Dupont, L.; Coquery, G.; Kriegel, K. Accelerated active ageing test on SiC JFETs power module with silver joining technology for high temperature application. *Microelectron. Reliab.* **2009**, *49*, 1375–1380. [CrossRef]
9. Scheuermann, U. Reliability challenges of automotive power electronics. *Microelectron. Reliab.* **2009**, *49*, 1319–1325. [CrossRef]
10. Amro, R.; Lutz, J.; Rudzki, J. Double-sided low-temperature joining technique for power cycling capability at high temperature. In Proceedings of the 44th European Conference on Power Electronics and Applications, Seville, Spain, 12–15 December 2005; Volume 10, p. 10.
11. Yu, H.; Li, L.; Zhang, Y. Silver nanoparticle-based thermal interface materials with ultra-low thermal resistance for power electronics applications. *Scr. Mater.* **2012**, *66*, 931–934. [CrossRef]
12. Siow, K.S. Mechanical properties of nano-silver joints as die attach materials. *J. Alloys Compd.* **2012**, *514*, 6–19. [CrossRef]
13. Bai, J.G.; Zhang, Z.Z.; Calata, J.N. Low-temperature sintered nanoscale silver as a novel semiconductor device-metallized substrate interconnect material. *IEEE Trans. Compon. Packag. Technol.* **2006**, *29*, 589–593. [CrossRef]
14. Ogura, H.; Maruyama, M.; Matsubayashi, R. Carboxylate-passivated silver nanoparticles and their application to sintered interconnection: A replacement for high temperature lead-rich solders. *J. Electron. Mater.* **2010**, *39*, 1233–1240. [CrossRef]

15. Li, X.; Chen, G.; Wang, L. Creep properties of low-temperature sintered nano-silver lap shear joints. *Mater. Sci. Eng. A* **2013**, *579*, 108–113. [CrossRef]

16. Ide, E.; Angata, S.; Hirose, A. Metal-metal bonding process using Ag metallo-organic nanoparticles. *Acta Mater.* **2005**, *53*, 2385–2393. [CrossRef]

17. Li, X.; Chen, G.; Chen, X. Mechanical property evaluation of nano-silver paste sintered joint using lap-shear test. *Solder. Surf. Mt. Technol.* **2012**, *24*, 120–126. [CrossRef]

18. Albert, A.D.; Becker, M.F.; Keto, J.W. Low temperature, pressure-assisted sintering of nanoparticulate silver films. *Acta Mater.* **2008**, *56*, 1820–1829. [CrossRef]

19. Roy, R.; Agrawal, D.; Cheng, J.; Gedevanishvili, S. Full sintering of powdered-metal bodies in a microwave field. *Nature* **1999**, *399*, 668–670.

20. Agarwala, M.; Bourell, D.; Beaman, J.; Marcus, H.; Barlow, J. Direct selective laser sintering of metals. *Rapid Prototyp. J.* **1995**, *1*, 26–36. [CrossRef]

21. Chaim, R. Liquid Film Capillary Mechanism for Densification of Ceramic Powders during Flash Sintering. *Materials* **2016**, *9*, 280–288. [CrossRef]

22. Wei, X.; Back, C.; Izhvanov, O.; Khasanov, O.L.; Haines, C.D.; Olevsky, E.A. Spark Plasma Sintering of Commercial Zirconium Carbide Powders: Densification Behavior and Mechanical Properties. *Materials* **2015**, *8*, 6043–6061. [CrossRef]

23. Lu, G.Q.; Li, W.L.; Mei, Y.H. Characterizations of Nanosilver Joints by Rapid Sintering at Low Temperature for Power Electronic Packaging. *IEEE Trans. Device Mater. Reliab.* **2014**, *14*, 623–629.

24. Suzuki, Y.; Ogura, T.; Takahashi, M. Low-current resistance spot welding of pure copper using silver oxide paste. *Mater. Charact.* **2014**, *98*, 186–192. [CrossRef]

25. Allen, M.L.; Aronniemi, M.; Mattila, T.; Alastalo, A.; Ojanperä, K.; Suhonen, M. Electrical sintering of nanoparticle structures. *Nanotechnology* **2008**, *19*, 175201. [CrossRef] [PubMed]

26. Lopes, W.A. Nonequilibrium self-assembly of metals on dibloCk copolymer templates. *Phys. Rev. E* **2002**, *65*, 031606. [CrossRef] [PubMed]

27. Bai, J.G.; Lei, T.G.; Calata, J.N. Control of nanosilver sintering attained through organic binder burnout. *J. Mater. Res.* **2007**, *22*, 3494. [CrossRef]

28. Li, Z.L.; Becker, F.; Stoll, M.P. Evaluation of six methods for extracting relative emissivity spectra from thermal infrared images. *Remote Sens. Environ.* **1999**, *69*, 197–214. [CrossRef]

29. Lei, T.G.; Calata, J.N.; Lu, G.Q. Low-temperature sintering of nanoscale silver paste for attaching large-area chips. *IEEE Trans. Compon. Packag. Technol.* **2010**, *33*, 98–104. [CrossRef]

30. Yang, L.Y.; Odendaal, W.G.H. Measurement-based method to characterize parasitic parameters of the integrated power electronics modules. *IEEE Trans. Power Electron.* **2007**, *22*, 54–62. [CrossRef]

31. Wang, M.Y.; Lu, G.Q.; Mei, Y.H. Electrical method to measure the transient thermal impedance of insulated gate bipolar transistor module. *IEEE Trans. Power Electron.* **2015**, *8*, 1009–1016. [CrossRef]

32. Chung, S.Y.; Han, T.S.; Kim, S.Y. Reconstruction and evaluation of the air permeability of a cement paste specimen with a void distribution gradient using CT images and numerical methods. *Constr. Build. Mater.* **2015**, *87*, 45–53. [CrossRef]

33. Xiao, K.; Calata, J.N.; Zheng, H. Simplification of the nanosilver sintering process for large-area semiconductor chip bonding: Reduction of hot-pressing temperature below 200/spl deg/C. *IEEE Trans. Compon. Packag. Manuf. Technol.* **2013**, *3*, 1271–1278. [CrossRef]

34. Yang, S.; Xiang, D.; Bryant, A. Condition monitoring for device reliability in power electronic converters: A review. *IEEE Trans. Power Electron.* **2010**, *25*, 2734–2752. [CrossRef]

35. Navarro, L.A.; Perpina, X.; Godignon, P. Thermomechanical assessment of die-attach materials for wide bandgap semiconductor Device and harsh environment applications. *IEEE Trans. Power Electron.* **2014**, *29*, 2261–2271. [CrossRef]

36. Akada, Y.; Tatsumi, H.; Yamaguchi, T. Interfacial bonding mechanism using silver metallo-organic nanoparticles to bulk metals and observation of sintering behavior. *Mater. Trans.* **2008**, *49*, 1537–1545. [CrossRef]

37. Grouchko, M.; Popov, I.; Uvarov, V. Coalescence of silver nanoparticles at room temperature: Unusual crystal structure transformation and dendrite formation induced by self-assembly. *Langmuir* **2009**, *25*, 2501–2503. [CrossRef] [PubMed]

38. Wang, S.; Li, M.; Ji, H. Rapid pressureless low-temperature sintering of Ag nanoparticles for high-power density electronic packaging. *Scr. Mater.* **2013**, *69*, 789–792. [CrossRef]

39. *SIGC32T120R3E IGBT3 Power Chip, Datasheet*; Infineon Semiconductors Ltd.: Munich, Germany, 2014.

40. Rabkowski, J.; Tolstoy, G.; Peftitsis, D. Low-loss high-performance base-drive unit for SiC BJTs. *IEEE Trans. Power Electron.* **2012**, *27*, 2633–2643. [CrossRef]

41. Guha, A.; Narayanan, G. An improved dead-time compensation scheme for voltage source inverters considering the devIce switching transition times. In Proceedings of the IEEE 6th India International Conference on Power Electronics (IICPE), Johor Bahru, Malaysia, 13–14 October 2014; pp. 1–6.

42. MMG25H120XB6TN, Datasheet, MacMic Co., Ltd. Available online: http://www.macmicst.com/searchlist. asp?id=507&sortid=89 (accessed on 7 May 2016).

43. Egelkraut, S.; Frey, L.; Knoerr, M. Evolution of shear strength and microstructure of die bonding technologies for high temperature applications during thermal aging. In Proceedings of the 12th Electronics Packaging Technology Conference (EPTC), Singapore, 8–10 December 2010; pp. 660–667.

44. Chen, G.; Cao, Y.J.; Mei, Y.H. Pressure-Assisted Low-Temperature Sintering of Nanosilver Paste for 5 × 5-Chip Attachment. *IEEE Trans. Compon. Packag. Manuf. Technol.* **2012**, *2*, 1759–1767. [CrossRef]

45. Katsis, D.C.; Van Wyk, J.D. Void-induced thermal impedance in power semiconductor modules: Some transient temperature effects. *IEEE Trans. Ind. Appl.* **2003**, *39*, 1239–1246. [CrossRef]

46. Zhao, Y.; Wu, Y.; Evans, K. Evaluation of Ag sintering die attach for high temperature power module applications. In Proceedings of the 15th International Conference on Electronic Packaging Technology (ICEPT), Chengdu, China, 12–15 August 2014; pp. 200–204.

47. Kim, I.; Song, Y.A.; Jung, H.C. Effect of microstructural development on mechanical and electrical properties of inkjet-printed Ag films. *J. Electron. Mater.* **2008**, *37*, 1863–1868. [CrossRef]

48. Fu, S.C.; Mei, Y.H.; Lu, G.Q.; Li, X.; Chen, G.; Chen, X. Pressureless sintering of nanosilver paste at low temperature to join large area ($\geq$100 mm^2) power chips for electronic packaging. *Mater. Lett.* **2014**, *128*, 42–45. [CrossRef]

49. Polasik, S.J.; Williams, J.J.; Chawla, N. Fatigue crack initiation and propagation of binder-treated powder metallurgy steels. *Metall. Mater. Trans. A* **2002**, *33*, 73–81. [CrossRef]

50. Mourad, H.M.; Garikipati, K. Advances in the numerical treatment of grain-boundary migration: Coupling with mass transport and mechanics. *Comput. Methods Appl. Mech. Eng.* **2006**, *196*, 595–607. [CrossRef]

51. Pande, C.S.; Cooper, K.P. Nanomechanics of Hall-Petch relationship in nanocrystalline materials. *Prog. Mater. Sci.* **2009**, *54*, 689–706. [CrossRef]

52. Hanlon, T.; Kwon, Y.N.; Suresh, S. Grain size effects on the fatigue response of nanocrystalline metals. *Scr. Mater.* **2003**, *49*, 675–680. [CrossRef]

MDPI AG

St. Alban-Anlage 66

4052 Basel, Switzerland

Tel. +41 61 683 77 34

Fax +41 61 302 89 18

http://www.mdpi.com

Materials Editorial Office

E-mail: materials@mdpi.com

http://www.mdpi.com/journal/materials

www.ingramcontent.com/pod-product-compliance
Lightning Source LLC
Chambersburg PA
CBHW051854210326
41597CB00033B/5894